From Construction to Re-construction
Heterogeneous Isomorphism Campus Architecture of Hunan University

从营造到建构
—— 湖南大学校园建筑

魏春雨　宋明星　著

中国建筑工业出版社
CHINA ARCHITECTURE & BUILDING PRESS

图书在版编目（CIP）数据

从营造到建构：湖南大学校园建筑 / 魏春雨，宋明星著. —北京：中国建筑工业出版社，2019.11
ISBN 978-7-112-13864-7

Ⅰ. ①从… Ⅱ. ①魏… ②宋… Ⅲ. ①湖南大学—教育建筑—建筑史 Ⅳ. ①TU244.3-092

中国版本图书馆CIP数据核字(2019)第242546号

责任编辑：陈 桦 王 惠
装帧设计：罗 漾
责任校对：芦欣甜
摄　　影：高雪雪 许昊皓 姚 力 胡 骉
翻　　译：谢 菲

从营造到建构——湖南大学校园建筑
魏春雨　宋明星　著

中国建筑工业出版社出版、发行（北京海淀三里河路9号）
各地新华书店、建筑书店经销
湖南省新创印务有限公司印刷

*

开本：889×1194毫米　1/12　印张：15 2/3　字数：341千字
2019年11月第一版　　2019年11月第一次印刷
定价：**79.00元**
ISBN 978-7-112-13864-7
（34927）

版权所有　翻印必究
如有印装质量问题，可寄本社退换
（邮政编码100037）

谨以此书纪念刘敦桢、柳士英先生创建湖南大学建筑学科 90 周年，感谢为她的发展而努力工作的人们。

This book is dedicated to 90 anniversary of Architecture discipline establishment in Hunan University, the discipline initiators, Mr. Liu Dunzhen and Mr. Liu Shiying, and the tremendous contributions of people to its development.

目录
Content

引言 **Preface**	14
湖南大学校园及建筑的发展历程 Heterogeneous Isomorphism of Campus Architecture	16
书院篇 **The Academy Architecture**	18
"异质同构"下的近代岳麓书院 Yuelu Academy and the Campus	24
吹香亭 Chuixiang Pavilion	32
风雩亭 Fengyu Pavilion	32
自卑亭 Zibei Pavilion	33
岳麓书院建筑群修复名录 Building Repair List of Yuelu Academy	34
先贤篇 **The Influential Predecessors**	35
柳士英 Liu Shiying	38
二院 The Teaching and Research Centre No.2	42
原图书馆 The Grand Library	48
二舍 The Student Hall No.2	52
科学馆 The Building of Science and Technology	58
九舍 The Student Hall No.9	70
七舍 The Student Hall No.7	78
胜利斋 The Shengli Residence Hall	86
原六舍 The Former Student Hall No.6	92
工程馆 Building for the Department of Engineering	98
大礼堂 The Grand Hall of Hunan University	108
图书馆 The Library	124

营造篇　The Campus Architecture ... 137

书院博物馆
The Museum of Chinese Academy ... 139

生物技术综合楼
Building of Bio-science ... 144

复临舍教学楼
Building of Fulin ... 145

法学院、建筑学院建筑群
The Building Complex for the School of Law and the School of Architecture ... 147

工商管理学院楼
Building of the Business School ... 150

软件学院楼
Building of the School of Computer Science ... 152

综合教学楼
The Building Complex for Education ... 154

游泳馆
Aquatic Center ... 156

研究生院楼
Graduate School ... 158

理工教学楼
Science and Engineering Building ... 160

异质同构的校园　Creating a Heterogeneous Isomorphic Campus ... 162

校园建筑的异质与同构
Heterogeneous Isomorphism of Campus Architecture ... 164

"异质同构"的校园建筑设计手法
Architectural Design Approaches of Heterogeneous Isomorphic Campus ... 166

场所的语义：从功能关系到结构关系
—— 湖南大学天马新校区规划与建筑设计 ... 170

附录　Appendix ... 180

湖南大学校园重要建筑一览表 ... 180

湖南大学的今昔 ... 183

后记·致谢　Epilogue & Acknowledgement ... 185

湖南大学校园（2005）

湖南大学校园 (2019)

引言
Preface

湖南大学的前身是创建于北宋公元 976 年的岳麓书院，历经宋、元、明、清等朝代的时势变迁，一直保持着文化教育的连续性。1903 年岳麓书院改制为湖南高等学堂，1926 年定名湖南大学，1937 年成为国民政府教育部十余所国立大学之一。

千余年来，学脉延绵，弦歌不绝，故有"千年学府"之称。"纳于大麓、藏之名山"是湖南大学校园的总体形态特征，校园西倚岳麓山，东望橘子洲及长沙城，北有凤凰山，南抵天马山，处于山水洲城之核心。校园自湘江西岸从古牌楼口到校园自卑亭，至岳麓书院及清风峡，形成一条千年的古文脉。校园沿山麓逶迤展开，与城市有机融合，形成特有的开放型校园。近年来校园内新建、改建、扩建了一批建筑，对建筑的原有功能进行了整合，各个区域之间既相互独立又相互联系，并始终保持其内在活力。各个历史时期不同风格的建筑尊重山水格局，注重尺度控制，注意对自然质感的材料与共生原则的运用，共同组成了多元化且独具特色的校园风景，"异质"却"同构"。

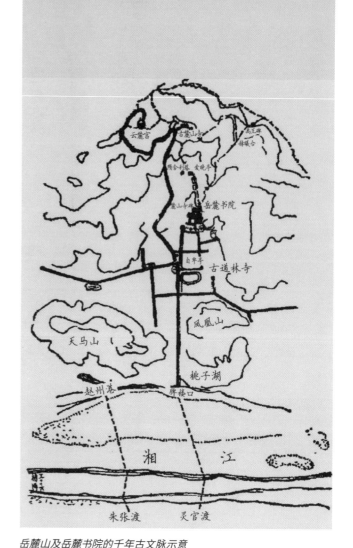

岳麓山及岳麓书院的千年古文脉示意
Schematic diagram of historic context for the Yuelu Mountain and Yuelu Academy

Hunan University was originated from Yuelu Academy, established in 976 AD in the Northern Song Dynast. Yuelu Academy went through the Song, Yuan, Ming and Qing dynasty and now has still held the continuity of culture and education. In 1903 Yuelu Academy was transformed and called 'Hunan College of Higher Education'. In 1926, the name was changed to 'Hunan University', and it was listed in the ten national universities managed by the Ministry of Education of the National Government in 1937. With over thousand years of education history, Hunan University enjoys the reputation of "Millennium university with centuries of fame".

"The famous mountain of scenic beauty, an ideal cradle-land of talents". The location of the Hunan University campus gives it unique possibilities: Yuelu Mountain sits at the west of the campus, a historical setting for discussion on succession of history; Tianma valley at the south stretches for miles along the waterfront of Xiang river, a lively space for accommodation and resident housing; Phoenix hill at the north side connects to another campus next door and provides an open access to the city, a natural symbol for the education ethos of Hunan University, and finally, the waterfront area at the east offers an open view to the hustle-bustle of central city lying to the east bank of the river. Over years of redevelopment in the campus, the university has kept its identity by creating a modern and open space for the education needs while at the same time having respect for the past. Planning of the campus places emphasis on landscape layout, building-scale control, natural materials utilization, and environmental harmony. The university campus generally features pluralistic culture and picturesque scenes and as results of redevelopment, it emerges 'Heterogeneous' in the 'Isomorphism' context.

岳麓山、岳麓书院和湖南大学建筑群形成的轴线

An axis formed by the Yuelu Mountain, the Yuelu Academy, and the campus of Hunan University

湖南大学校园及建筑的发展历程

Heterogeneous Isomorphism of Campus Architecture

湖南大学校园的发展历程可以分为三个时期，第一个时期是传统岳麓书院时期，学校的主体建筑为岳麓书院建筑群。第二个时期是从 1926 年更名为湖南大学后，由刘敦桢、柳士英主持规划进行建设的理性开放的营造时期，这一时期，湖南大学建设了一批高质量的校园建筑，如湖大二院、科学馆、工程馆、大礼堂等。第三个时期则是改革开放后，进入新地域校园的建构时期，尤其 2003 年后，学校在天马西麓征地 200 余亩进行新校区的建设，又建设了一批校园建筑，如综合教学楼、软件大楼、研究生院楼、理工楼、游泳馆、超算中心等。至此，湖南大学的总体格局基本成型。

There are three distinguish building types developed along the history of the Hunan University campus. First, academy architecture started thousand years ago and is still in the process of development. Second, modern architecture spread widely in the campus when Chinese pioneers of modern architecture, i.e. Mr. Liu Dunzhen, Liu Shiying , came the university to take charge of the campus construction after 1926. In the third stage of transformation, the new regionalism architecture becomes dominant in the nowadays modern campus. Particularly after 2003, about 13.33 hectares of land had been expropriated and designated as a new development site upon which the complex buildings for Education, the buildings for school of Information Science, the graduate school, building complex for Engineering, indoor swimming pool, National Center for Supercomputing Applications and son on were built in the last few years. The spatial structure of the campus hence started to take shape and at the same time the University entered a new era of development.

书院篇 诗意的原型
（岳麓书院时期）公元 976 — 1926 年
The Academy Architecture A Poetic Prototype
(Yuelu Academy) AD 976-1926

作为我国古代四大书院之一，岳麓书院前身可追溯到唐末五代（约 958 年）智睿等二僧办学时期。北宋开宝九年（976 年），潭州太守朱洞在僧人办学的基础上，正式创立岳麓书院。嗣后，历经宋、元、明、清各代，至清末光绪二十九年（1903 年）改为湖南高等学堂，尔后相继改为湖南高等师范学校、湖南工业专门学校，1926 年正式定名为湖南大学。

Yuelu Academy is one of the four ancient Chinese academies of classic learning. Its history could be traced back to a school run by two monks in the late Tang Dynasty (around AD 958). In 976, Yuelu Academy was officially funded by Mr. Zhu Dong, magistrate of Tanzhou prefecture. It survived the dynasty alternation of centuries, such as Song, Yuan, Ming and Qing. In 1903, at the late of Qing dynasty, it functioned as a school for higher learning called "Hunan Da Xuetang", and then changed to Hunan Normal College, Hunan Public Polytechnic School in succession. Finally in 1926 it took its current name "Hunan Daxue (Hunan University)".

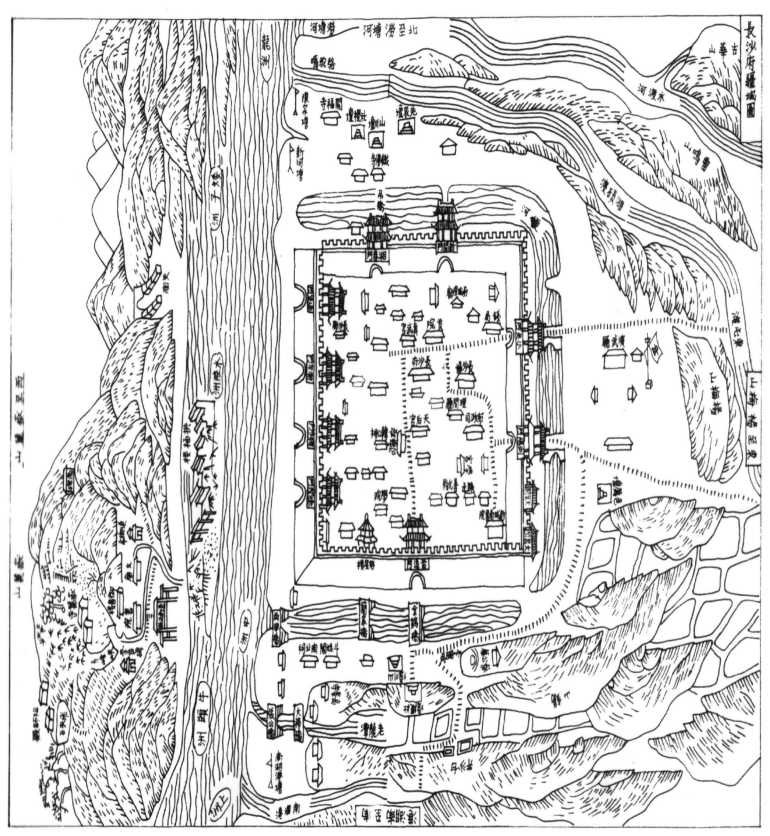

长沙府疆域图 清乾隆十四年《长沙府志》

中国古代书院选址通常并不择于城中，而是避开城市之喧嚣，择清净山野风水俱佳之地，尤其注重自然山水景致对读书人的影响，以及在山野僻静之地思考家国天下之事，感悟儒家理学经典的至理，"寓情于景，情景交融，寓意于物，以物比德"。岳麓书院顺着岳麓山东麓清风峡山谷而建，隔湘江与长沙老城相望。现存建筑大部分为明清遗物，其规划格局一直未变延续而来，自大门、二门、讲堂、御书楼的中轴线，半学斋、教学斋、百泉轩、湘水校经堂、文庙、后花园居于中轴两侧，分为讲学、藏书、供祀、园林四大部分，各部分通过廊道和天井连为一体，整体为一座庭院式砖木结构建筑。书院空间秩序蕴含着儒家文化尊卑有序的哲理，强调社会伦理关系中的等级，反映在空间上就是主次空间分明，以中轴、侧向轴线、纵深多进的院落，表达庄严、理性但又幽远、静谧的环境氛围。整个书院建筑群较为完整地呈现了中国古代文人建筑避世、儒雅、悟道、严肃的气质。

Yuelu Academy is lying quietly in the valley of Breeze Cool at the east mountain foot, overlooking the city centre of Changsha to the east bank of Xiang River. Most of the existing building complex is relics of the Ming and Qing Dynasty, and it basically comprises four parts used for different purposes, such as lecturing, library, worship and leisure in the garden. Each part of the academy interconnects each other by courtyards as a whole, and it is a characteristic layout of a traditional Chinese brick-timber structure. The spatial sequence of the academy implies underlying social value and ethical relation in classic Confucian culture, which reflects a strict hierarchy system of a class society and concepts of superiority and inferiority. The building complex is symmetrically arranged in general and the buildings of significance including the Main Gate, the First Gate, the Second Gate, auditorium, study hall, lecture theatre, Baiquan pavilion, Yushu library, Xiang research institute and Confusion temple are positioned in order along the central axis. Other buildings for accommodation, worship etc. are built away from the central line according to the zoning strategies of the academy. The overall building plan reveals principles of Chinese traditional architecture: building siting has regard to the environment inhabitability ("Fengshu") upon which a harmony relationship between man and nature could be built, and favorable built environment may be therefore created to inspire occupants both physically and psychologically. Chinese academy architecture has the same principles and follows the motto of Confucian: "being sympathetic to nature for open-mind, being thoughtful to matter for self-integrity".

南宋乾道三年（公元1167年），朱熹与张栻在岳麓书院讲堂讲学，"论中庸之义，三昼夜不能合"，"道林三百众，书院一千徒"，"坐不能容"，"饮马池水立涸，与止冠冕塞途"，是为当时之盛况写照。这反映了中国传统书院的开放姿态，并不与世隔绝。

In 1167, Confucian master Zhu Xi and Zhang Shi gave open lectures in the auditorium of the Yuelu Academy. They presented views on different Confucian schools and debated openly on hot topics in Confucian. There was a record turnout of over thousands attendees for such an event of that time. It is a typical portrait of open education mode in the traditional Chinese Academy.

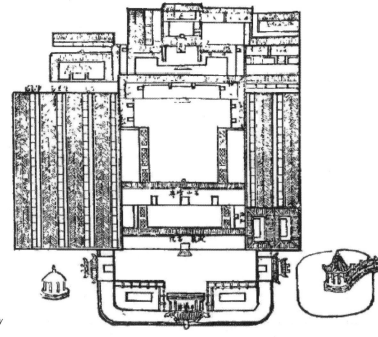

· 清代岳麓书院图
Historic map of Yule Academy

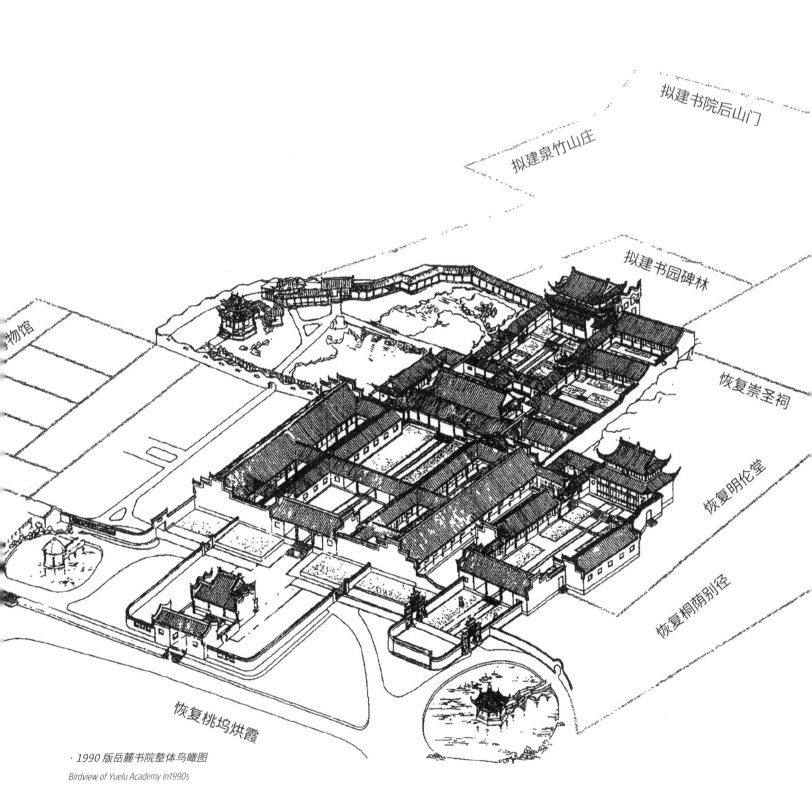

· 1990 版岳麓书院整体鸟瞰图

Birdview of Yuelu Academy in 1990s

"异质同构"下的近代岳麓书院
Yuelu Academy and the Campus

在 20 世纪 30 年代，湖南大学校园曾经一度以书院的建筑为主体，书院的传统建筑群仍然作为现代学校的校舍。这一时期的校园即岳麓书院给我们的启示包括以下几点。

Yuelu Academy was once dominant in the university campus and then functions as a part of educational infrastructure in the campus of the modern age. Yuelu Academy of this period suggests that it has the following valuable architectural aspects in light of campus development.

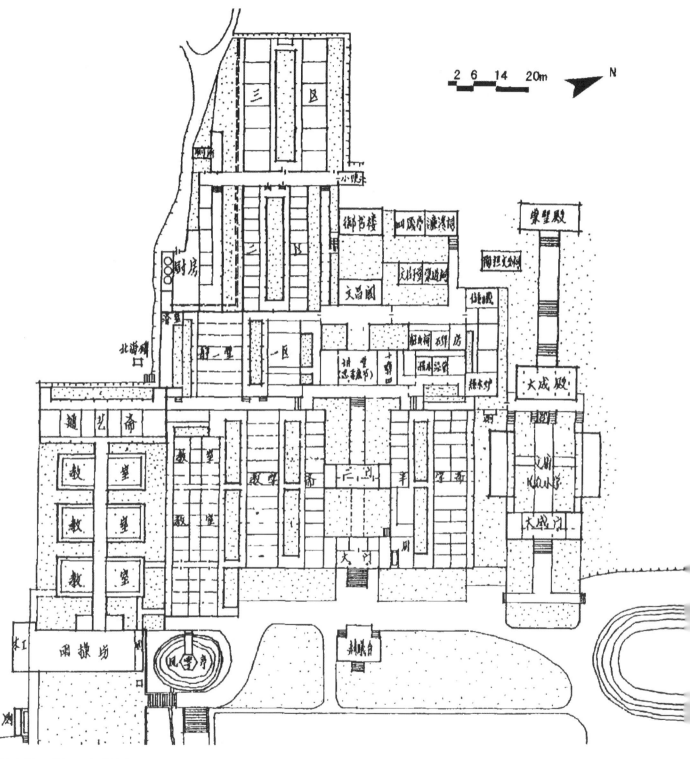

· 湖南大学20世纪30年代校舍平面
The Campus Map of Hunan University in 1930s

· 赫曦台　Hexi Pavilion

· 二门　The Second Gate

· 讲堂　The Auditorium

· 碑廊　The Stele Gallery

1) 校园与城市虽相对分离，但其办学开放且兼容并蓄。

2) 高度整合与复合化的空间。教读的讲堂与学习的斋舍、游憩空间完全整合为一体，其功能的融合性之高，往往超越现代所谓机械式的功能分区的校园。这种空间的互动与相互映衬，使它的使用效能优于现代简单的泾渭分明的专属空间校园布局。

1) Academy was though located in a remote region from city, but its education system pursued openness and excellence in integration.

2) Spatial arrangement of academy was highly integrative and compact. One building complex could facilitate purposes of teaching, study, research and recreation. The space organization appears particularly efficient and is better than the functional zoning in modern buildings.

3) 适应地域气候特征。岳麓书院建筑群以基本的"庭院"单元为原型：天井相间，屋脊相连，形成连绵如蜂窝状、四通八达的整体建筑群。"行至幽厢疑抵壁，推门又见一重庭"，具有湖南传统聚落的典型特征。

3) Building layout responded to bio-climate of the region. Courtyard is the building prototype of the academy, which is advantageous to create a favorable environment inside as well as to allow buildings to inter-connect each other at any direction according to space requirement. Its overall configuration turns out " passage seems never to end as one courtyard after another".

· 后花园 · 庭院 1 · 庭院 2	· 庭院 4	Courtyard 4
The Back-yard Garden Courtyard 1 Courtyard 2	· 庭院 5	Courtyard 5
· 庭院 3 Courtyard 3	· 庭院 6	Courtyard 6

· 书院山墙
Gables of Yuelu Academy

· 庭院 7
Courtyard 7

4）书院体现了尊卑有序、伦理有别的伦理关系，并且有清晰的轴线，但不拘泥于轴线，其空间在局部可依循环境关系发生偏转。

5）书院多为一组较为庞大、严谨、规整的建筑群，但由于重视地形的利用，多依山而建，层层叠进，错落有致，建筑虽较封闭但不框囿于形，而环境的开拓则十分开敞，两者有机联系，收到"骨色相和，神彩互发"之效，这为校园后来的自由发展与布局埋下了伏笔。

4) Academy architecture follows classic principles in terms of spatial order, but it is also pragmatic and seems to put function at the first place when the surroundings are not flexible.

5) Academy architecture is a large scale of building complex. Its site is carefully selected and its building-form reflects the surrounding conditions. The focus of attention is generally given to hold a balance and harmony relation between man and nature. This set up a basic law for later campus development that is "to harmonize with surroundings, to synergize for mutual benefits".

· 景窗
Side windows

· 庭院 8
Courtyard 8

· 庭院 9
Courtyard 9

吹香亭
Chuixiang Pavilion

岳麓书院大门北侧百十步外，文庙院墙邻侧，有一池塘曰"黉（音 hong）门池"，据传凿于宋代。"黉"即古时候的学校。池上有亭，曰吹香亭，池中遍布荷花，"风荷晚香"（岳麓八景之一）即指此地。

Chuixiang Pavilion: the pavilion is located outside of the screen wall of the Confucian Temple, at the north side of the Main Gate. A pond called "Hongmen" surrounds the pavilion that was originally used for watering horses. Its history could be traced back to Song Dynasty 1000 years ago. In summer, this pond is covered with green leaves of water lily stippled with lotus blossom, and the air is full of flower fragrance. This beautiful scene has been well known as "Chuixiang pavilion in a lotus evening", which is one of eight famous scenes in the Yuelu Mountain area. The pavilion therefore took its name after it, "chuixiang pavilion".

风雩亭
Fengyu Pavilion

岳麓书院大门南侧百十步外，与吹香亭相对之处，有一池塘曰"饮马池"，原池据传凿于宋代。南宋大理学家朱熹来岳麓书院与张栻会讲，吸引了众多学者听讲，其所乘之马匹置于此塘饮水，将池塘的水都饮光了，后人据此将其命名为饮马池。池塘之上的草亭，嘉庆年间的院长欧阳厚均重修后改名为风雩亭，现亭为 1984 年重建。

A round horse-pond lies the south side of the Main Gate. This pond was constructed as early as Song Dynasty. In 1167, Confucian master Zhu Xi came to the Yuelu Academy and gave a series of open lectures on the doctrine of Confucius. The number of horses accompanying the attendees was so big that the water of the whole pond had been used out. Since then, the pond had been called "Horse-pond". The thatched pavilion in the pond was named as the West Pavilion. But after its renovation work led by the Academy President of the time, Mr.Ouyang Houjun in the Song Dynasty, the pavilion had its current name "Fengyu Pavilion", signifying the Pavilion was dedicated to the nature delight. The pavilion nowadays was rebuilt in 1984.

自卑亭
Zibei Pavilion

　　岳麓书院大门东侧往湘江方向200米,东方红广场东北角有一亭,即自卑亭。其位置在灵官渡、湘江、牌楼口、自卑亭、岳麓书院、清风峡的古轴线上,亭名取自《中庸》:"君子之道,譬如远行,必自迩;譬如登高,必自卑。"迩,近也;卑,低矣。意思是君子的道德修养之道,好比长途跋涉,必须从近处开始,又比如攀登高山,必须从低处开始。自卑亭前面是浪花涛涛的湘江,背面是郁郁葱葱的岳麓山,正如进入岳麓书院的前导空间。

　　Zibei Pavilion / the Humble Pavilion : the pavilion sits quietly at the north-east corner of the Dongfanghong Square, about 200 meters away to the Main Gate of the Yuelu Academy. It marked the entrance of the old academy district. Its name originates from the quotation in 'the Doctrine of the Mean':" the way to be a man of honor is a long trek made by a very step at near distance, and is a hard mountaineering expedition started from the place at the lowest.", namely, to be a great man should begin with a humble mind.

岳麓书院建筑群修复名录
Building Repair List of Yuelu Academy

建筑名称 (List of Projects)	修复时间 建筑设计师 (Restoration period, Architects)
大门、二门、讲堂、赫曦台 (Main Gate, Second Gate, Lecture Hall, Hexi Pavilion)	1982 — 1984 年 杨慎初、黄善言、陈自艾、陈颖初 1982-1984, Yang Shenchu, Huang Shanyan, Chen Ziai, Chen Yinchu
御书楼 (Yushu Library)	1984 — 1986 年 杨慎初、黄善言 1984-1986, Yang Shenchu, Huang Shanyan
教学斋、半学斋 (Lecture theatres, Study and Offices)	1984 — 1986 年 杨慎初、黄善言、陈自艾、汤羽扬 1984-1986, Yang Shenchu, Huang Shanyan, Chen Shengxin, Tang Yuyang
湘水校经堂、专祠 (Xiang research institute, Memorial temple)	1985 — 1986 年 杨慎初、黄善言、陈升信、汤羽扬 1985-1986, Yang Shenchu, Huang Shanyan, Chen Shengxin, Tang Yuyang
风雩亭、吹香亭、牌楼 (Fengyu Pavilion/Season-delight Pavilion, Chuixiang/Sweet-breeze Pavilion, Ceremonial Arch)	1985 — 1986 年 杨慎初、黄善言、陈升信、汤羽扬 1985-1986, Yang Shenchu, Huang Shanyan, Chen Shengxin, Tang Yuyang
百泉轩 (Baiquan Pavilion)	1985 — 1986 年 杨慎初、黄善言、陈升信 1985-1986, Yang Shenchu, Huang Shanyan, Chen Shengxin
文庙大成殿 (Great Hall of Confucius)	1986 — 1988 年 杨慎初、黄善言、魏春雨 1986-1988, Yang Shenchu, Huang Shanyan, Wei Chunyu
园林、后门、碑廊 (Garden, Rear Door, Stele Gallery)	1988 — 1991 年 杨慎初、柳肃 1988-1991, Yang Shenchu, Liu Su
麓山寺碑亭 (Stele Pavilion of Lushan Temple)	1989 — 1992 年 杨慎初、汤羽扬 1989-1992, Yang Shenchu, Tang Yuyang
文庙大成门、两厢 (Grand Gate of the Confucian Temple, Wing-rooms of the Confucian Temple)	1991 — 1993 年 杨慎初、柳肃 1991-1993, Yang Shenchu, Liu Su
时务轩 (Shiwu Memorial Pavilion)	1993 — 1994 年 杨慎初、张卫 1993-1994, Yang Shenchu, Zhang Wei
杉庵、后山门、泉竹山房 (Fir Woods Hut, Back Gate, Spring and Bamboo House)	1998 — 2000 年 蔡道馨 1998-2000, Cai Daoxin
文庙崇圣祠、明伦堂 (Chapel of the Confucian Temple, Minglun Lecture Hall)	2002 — 2004 年 柳肃 2002-2004, Liu Su
道中庸亭、极高明亭及连廊 (Juste-milieu Pavilion, Wisest Temple, Corridor)	2003 — 2004 年 柳肃 2003-2004, Liu Su
学习斋 (Study and Recidence Hall)	2003 — 2005 年 柳肃 2003-2005, Liu Su
书院博物馆 (Museum of Yuelu Academy)	2009 — 2012 年 魏春雨 齐靖 2009-2012, Wei Chunyu, Qi Jing

先贤篇 理性的开放
（早期湖南大学建立时期）1929—1953 年

The Influential Predecessors Rationalism and Openness
(Hunan University) 1929-1953

刘敦桢与柳士英为开创湖南大学建筑学科的两位先贤。1929 年，刘敦桢在湖南大学土木系中创立建筑组，是为湖南大学建筑学科的开端。随后刘敦桢应邀去北京参加中国营造学社的工作，邀请其留日师兄柳士英来湖南。柳士英 1934 年来到湖南大学，继续发展建筑学教育，聘请了蔡泽奉、许推等一批留日、留欧美的学者担任教授，此时有建筑学专业教师十余人，成为一支强大的建筑学专业队伍。1953 年院系调整，湖南大学、中山大学、南昌大学、广西大学、武汉大学、云南大学、四川大学的土建与道路、铁建各专业合并，成立中南土木建筑学院，成为中南地区实力最强的土木类的学院，柳士英任院长。从刘敦桢、柳士英、蔡泽奉等人在 20 世纪 20、30、40 年代设计校园建筑开始，到之后杨慎初、黄善言对岳麓书院的修复，直至今日的教学建筑、校园环境设计，湖南大学建筑学院始终进行着对湖南大学校园的设计、改造、保护工作。柳士英先生在湖南大学一直从事教学与设计工作直至 1973 年逝世，他不仅制定了这一时期湖南大学的校园规划，也在这里留下了大量经典的校园建筑作品。因此，湖南大学的校园被深深地打上了柳氏的烙印。

There were two important figures in the history of architectural education in Hunan University: Mr. Liu Dunzhen and Mr. Liu Shiying. In 1929 Mr. Liu Dunzhen set up architectural research group within the Department of Civil Engineering, which marked the beginning for architectural disciplinary development. Then he invited his alumnus Mr. Liu Shiying to join him in the university when he went to Beijing for the work of the Society for the Study of Chinese Architecture (SSCA). Since Mr. Liu Shiying came to Hunan University, more scholars with overseas study background had taken lecturing positions in the university. A team of over ten architectural professionals gradually took shape. In 1953, the departments of civil engineering in several key universities, such as Hunan University, Sun Yat-sen University, Nanchang University, Guangxi University, Wuhan University, Yunnan University, and Sichuan University, merged together and founded a School of Civil Engineering of the Middle South of China under the leadership of Mr. Liu Shiying. The work of construction on the university campus was initiated by Mr. Liu Dunzhen, Mr. Liu Shiying and Mr. Cai Zefeng in 1920s, 1930s, and 1940s respectively. Following the restoration of Yuelu Academy led by Mr. Yang Shenchu and Mr. Huang Shanyan, School of Architecture in the Hunan University has carried out work regarding planning, environment strategies, campus redevelopment and building renewal in the university campus ever since. Mr. Liu Shiying had engaged very work of architecture education and design in the campus until he died in 1973. A great number of architectural works he had done remains in the campus and become precious living memories of the history of this period.

岳麓书院自建立之初就形成了一条明显的中轴线，它自上而下经自卑亭、牌楼口、一直延伸到湘江边。这在规划中是非常重要的一条轴线，但柳士英力图避开这条轴线，以理性、开放的思想建构理想中的新型校园，但与此同时，对旧的文脉与形式却保持了最大的尊重。他根据校园的地形地貌特征提出四个同心圆的规划理念，核心的圆为体育活动区，第二个圆为教学区，第三个圆为学生活动区，第四个圆圈为教职工生活区。最外层为岳麓山景区。体育区在最中心方便师生集合，教职工生活区接近公园景区，成为师生课余休憩之地。日常教学、教职工、学生三个区之间相互融合，联系便捷。而美丽的岳麓山环绕校园四周，形成了以广场、运动场、绿地、大型公共教学楼为主的开放型校园。

The campus has a clear central axis starting from the academy, passing through Humble Pavilion and down to the waterfront of Xiang River. It was a reference line to guide the campus planning. However with respect to the tradition culture and exiting terrain situation, Mr. Liu Shiying proposed a new plan of the campus upon the axial mode: four concentric rings concept. The central area of the rings was designated to sport events. The area of the second ring facilitated space requirement for education, meanwhile the student accommodation and residential sector for staff located in the third and fourth ring respectively. At the outmost area was the resort of Yuelu Mountain located close to the staff living district. This plan allows different districts of the campus to stay independent but inter-connect efficiently when it is needed. Consequently an open campus took its form with a square, playgrounds, green-base and relative bulky educational buildings in the central area .

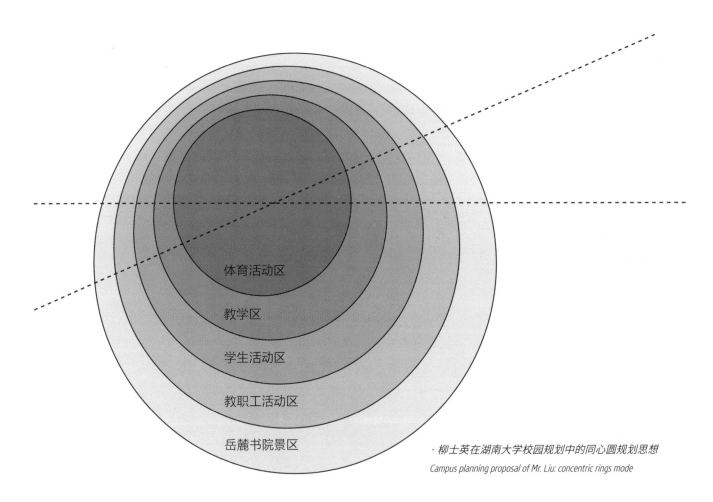

· 柳士英在湖南大学校园规划中的同心圆规划思想
Campus planning proposal of Mr. Liu: concentric rings mode

湖南大学在中国的大学校园中具有非同一般的独特性，那就是她的"开放性"。一直以来，湖大没有围墙，甚至没有校门，市政道路、公共交通及岳麓山的游客穿行相间。在这样的校园中，校内与校外，社会与大学的区别被模糊了，古与今的界限也被模糊了。

　　这一时期的校园与建筑变化给我们的启示是：

　　一是隐形的轴线。由于柳士英的规划，岳麓书院原有的强烈的轴线感被消解了，校园的结构更加自由，形成同心圆式的分散式布局，在校园的规划中体现了现代大学的自由精神。

　　二是形成了"异质同构"的校园特色。这一段时期建成的校园建筑不拘泥于传统形式，大胆创新，取各家之长，在校园内中国与西洋、传统与现代、大学与城市之间，形成了共生与融合的氛围。

　　这是校园的第一次传承与"变异"，是一次异质同构的过程。自此确立了兼容并蓄的开放型校园特质。

Hunan University features unique "openness" in the education system and campus organization. Until now, the university campus pride herself on holding an open attitude both to the public and the city. The campus interweaves the fabric of the Changsha city, hereby the boundary between the 'inside' and 'outside', the campus and community, eventually the past and present is unidentifiable.

The university campus of this period had notable characteristics listed as below:

1) Resilient structure of campus. There were two basic modes of planning interrelating with one another so that a free and open campus has been created. The present university campus could be an epitome of modern university of free spirit.

2) Campus architecture of heterogeneous isomorphism. Having respect for historical buildings, the predecessors of modern architecture in this period took innovative concepts of modern architecture and technologies with ambition to build a unique university campus for the locals.

To sum, the campus of Hunan University had established her uniqueness in openness and integration after its first transformation.

柳士英
Liu Shiying

中国建筑学教育先行者
湖南大学建筑学科创建人

柳士英（1893—1973）

　　柳士英（汉，1893.11—1973.7，苏州人）建筑学家，建筑教育家。早年留学日本，1920年毕业于东京高等工业学校建筑科。1922年在上海与刘敦桢等组建"华海建筑师事务所"。1923年在苏州工业专门学校创办建筑科，为我国近代建筑教育之开端。1953年筹建饮誉海内外的中南土木建筑学院。柳士英沉静笃厚，不趋名利；作品清新雅朴，融汇中西。

一、华海建筑师事务所

柳士英1920年毕业回国到了上海,正值工业资本扩张。他在日华纱厂做了一年建厂施工员,后转东亚公司,随后转日本冈野建筑师事务所,任设计师。1922年,柳士英毅然脱离冈野,与留日同学兼挚友刘敦桢、王克生、朱士圭一道,创设了"华海建筑师事务所"。所址设于上海九江路与江西路口,独立自营设计业务。从时间上看,"华海"比目前认为的国人最早的事务所"庄俊建筑师事务所",要早两年,应为最早。

但"华海"未有壮大发展,一是由于外国洋行打样间倾轧压制;二是当时民族资本不敌外资,国人事务所少而无影响力,业务难觅;三是无上层政治背景,无财阀后盾。经过几年支撑,终因社会所迫,不得不停业中断。

1930年,柳士英回上海重理"华海"业务,以图重振。

除了"华海"设计项目外,柳士英还作了不少规划设计。1928年,当时苏州设市,柳士英被委任为苏州市政工程筹备处总工程师,1929年又任苏州工务局长,主持城市规划和营建。他整理旧有街道及河道系统,规划三道交通循环线。但这些改造措施影响了私人房地产得失,触犯了地方权贵恶势力的既得利益,阻力横生,进展迟缓,最终仅完成十分之一二。

二、苏工建筑科

1923年,"华海"业务受挫,难以施展,柳士英回到家乡苏州,利用苏州工业专门学校为基地,创立建筑科。苏州工业专门学校建筑科的创办真正揭开了我国近代正统建筑教育的历史。

柳士英自任建筑科主任,教授建筑营造、建筑设计、建筑史等主干课程。相邀刘敦桢、朱士圭协助创办,并来此执教。建筑科教学体制参承日本,偏重工程技术掌握及运用,以培养全面懂得建筑设计及施工的人才为目标。学制三年。开设建筑意匠(即建筑设计)、中西营造法、建筑历史、结构设计、测量、美术等课程。从专业课程设置上可窥见柳士英强调学以致用的理念。至1927年,毕业了两届学生。

1927年建筑科并入南京东南大学,嗣改国立第四中山大学,又改国立中央大学,就此成为中大建筑系之前身。这亦是我国大学建筑系之首创。正因如此,苏州建筑科值得书上一笔。

三、开创湖南建筑教育——中南土建学院

1934年,由刘敦桢力荐,当时湖南大学土木系主任唐艺菁亲赴上海相邀,柳士英来湘,任教于湖大土木系。从此致力于湖南,传播建筑教育,未曾离走。

柳士英是较早扎根内地传播建筑教育,从事设计实践的建筑师。1934年他设计的长沙电灯公司厂房及配套设施,为湖南最早的现代工业建筑之一。此外,他在长沙还设计了湘鄂赣粤四省物品展览会馆、商务印书馆、长沙上海银行、长沙医院、李文玉金号等建筑,并担任过长沙迪新土木建筑公司建筑师。

1934年,柳士英在湖大创设了建筑组,开创了湖南现代建筑教育。抗战前后又兼任长沙高等工业学校教授,辅佐省立克强学院开设建筑系,兼任系主任、教授。新中国成立后被任命为湖大土木系主任。1951年举办"建筑专科班"和"土木专科班",快速输送人才到第一个五年计划建设事业上。1952年主持完成了有湖南大学土木系和华南工学院建筑系参加的,武汉华中科技大学建校总体规划和多项学校建筑设计。

1953年院系调整,国务院决定合并湖南大学、武汉大学、南昌大学、广西大学、四川大学、云南大学、华南工学院所属土木系和铁道建筑专业,成立中南土木建筑学院,柳士英任筹备委员会主任。同年10月16日,中南土木建筑学院正式成立,柳士英任院长,设有营造建筑、道路建筑等系。《中南土建》杂志及《中南土木建筑学院学报》陆续发刊。短短几年,学院师资、科研、基建成绩斐然,成为国家土建人才的重要培养基地。1958年学院更名湖南工学院,柳士英任院长。1959年又改湖南大学,柳士英任副校长。

柳士英注重自身修养，认为建筑有社会性，要求设计者清廉完善，这是从事设计的第一步。他推崇朱熹在岳麓书院的训诫："循序渐进，熟读精思，虚心涵泳，切己体察，着紧用力，居敬持志。"习学建筑，亦可循学、问、思、辨之序，把直接经验加以系统化，如是方能革新学术，磨砺品行。

他在教学中尤为强调两方面：第一，须精通历史本原及发展。"无源之水，不能广矣"，认为是否了解历史实为内行与外行之重大区别。他对中西学术渊源理解颇深，加之本身存养省察，故能熟谙运用。第二，须熟悉技术构造处理，这是取得深厚广博造诣之必要条件。对重要的结构处理，他常亲自计算出图，如湖大礼堂舞台结构就是他完成出图的。他时常满堂讲透一个节点大样，并且认为"三分匠，七分主"，不求一锤定音，注意在施工中观察修改。

柳士英宽厚雅朴，不求闻达，工作一丝不苟，防杜敷衍，染濡师生。对组织无有烦劳奢求，年岁渐高，仍伏案精绘，孜孜不怠。教学过程中，他编著有建筑设计、建筑构造、建筑历史、结构等课程教材讲义，如《西洋建筑史》《五柱规范》《建筑营造学》。1959年发表"建筑美"一文。1962年开始培养建筑学研究生，为我国为数不多的培养建筑研究生最早的导师之一。同年担任原建工部教材编审委员会委员。

柳士英在湘40年，在湖大及中南土木建筑学院等院校传播建筑教育的同时，亦留下不少作品。原湖大校址多设于岳麓书院原址。1926年刘敦桢设计了湖大第二院。1933—1935年间，蔡泽奉设计的图书馆及科学馆相继落成。1934年柳士英对湖大校区进行了多处修葺、扩建和规划，对教学区、风景名胜、宿舍、实习工厂统筹安排，使"山灵虽奇，得人文而显"，始奠定了湖大校园之良好雏形。1938—1945年间，湖大多次遭受日军狂轰滥炸，图书馆及许多校舍被焚毁。柳士英不屈强暴，撤至湘西辰谿龙头建分校，因陋就简，修建了大量校舍。1946年复原迁还长沙本校，面对残垣断壁，柳士英组织拆迁修复、重建校园，设计建造了不少校舍，完成了科学馆改建加层，使欧式锁心石门及檐壁线脚与中式装饰及屋顶糅合，取得中西合璧的效果。是年，他又设计并招商承建新图书馆，对民族形式

进行了有益尝试，新图书馆体量顿隙有致，横侧高低仰借叠伏，处理较成熟，而无明显折中印迹。1948年柳士英设计的工程馆是具浓郁"机器美学"色彩的作品，以水平窗带之圆转隐喻机械皮带轮的上下传动，形象简洁实朴，既有学府味又回避了当时一般教学建筑呆板和抑郁之陋端，属上乘之作。1951年柳士英设计了湖大礼堂，当时湖大校长李达未经教育部批准即拍板上马。20世纪50年代伊始，苏联影响还未扩大，但已出现了"民族形式和社会主义内容"的前奏，大礼堂自然采用大屋顶形式，正面曲脊飞檐，交织错叠，侧面壁比扶摇壮观。值得注意的是檐下斗栱处理不恪守法则，似从檐下反挂而下，形成奇特装饰效果。教育部领导视察后大加赞扬，收回了对先斩后奏而发的责难，建议推广。从他20世纪50年代中期的许多作品中，至今依稀可见从"民族形式"到"反浪费"运动的偏摆痕迹。

四、设计观

柳士英留日求学间，正值各路早期现代思潮涌入日本之际，折中主义"洋风"、新艺术运动、分离派、表现派等纷至沓来，他接受了早期现代思想，主张合理的功能布局、协调的形体、严谨的比例尺度，从中窥悉到这些熏染。

对20世纪30年代的"中国固有形式"及"国粹主义"，他为之恻动，始终没有走完全照搬宫殿型固有形式的套路，是一个较具风骨，有主见的建筑师。他竭力推举现代思想，早期作品讲求结构逻辑，主张建筑应通过纯洁手法反映现实，以适应钢筋水泥材料特点。随着设计阅历加深，他的"自然主义"倾向愈加明显，清新雅朴的风格日臻成熟。虽然"功能至上，以真为美"的思想至深，但他不是一个彻头彻尾的功能主义者，其独特之处在于其作品中多残留有"新艺术运动"之遗风，喜好斟酌装饰与细节，踟蹰其间，寻得慰藉，取得巧妙的平衡。

柳士英对改良派的"中西合用，观其会通"主张持赞同意见，但对"中体西用"不以为然，认为不必强分主次尊卑，而应平等视之，态度开明，并作了一些颇有创意的探索。虽然最终亦未跳出"折中"围框，但可贵的是他能灵活处理，融会贯通，

决不恪守古典法式则例，不牵强附赘，即使20世纪50年代的"民族形式"作品，仍似清荷一枝，独具风范。其晚期设计中有不少值得借鉴之处，可用"细腻、变通、涵忍"加以既括：

细腻

细腻情寓细节之中，没有细节就没有建筑，缺乏细节的建筑是没有灵魂的，细部是建筑的浓缩。柳士英认为现代建筑的装饰在形式上与传统建筑不同，更需要在加工和材料使用方面格外审慎。细节、小尺度的部件处理能够弥补建筑视觉上的脱节和贫拮，尤擅如下处理。

曲线

他常把曲线糅进建筑，承前启后、上下过渡，无论由直转曲作"收头"，还是半圆转柱、弧形体、屋脊曲线，甚至线脚变化都能别赋情趣，各臻其妙。

收头

他有一个设计准则：凡事应有交代。线、面、体之过渡搭接要交代清晰。建筑亦有"起、承、仰、合、停"节奏变化，要总览斟酌。其收头处理体现在多方面：竖向段落常承沿"三段式"经典，灵活运用材料线脚变化来显示顶部收头处理；横向展延到一定时或停或转，如湖大工程馆西端以弧形体加折线体为收头，东端跃落一展作收头。细部收头处理更必不可缺。圆是最常用收头方式，戏称之曰"灵魂出窍"。收头要水至渠成，不可有矫揉之嫌。

光影

运用阳光是最经济实惠的装饰手段，他常把大片连续墙体作轻微波折形处理，以承接阳光，使形影相得益彰，迤逦波动。

入口

他认为追求形体奇异繁缛不应以牺牲入口形象、地位为代价，在建筑入口处下足功夫，决不会白费力气，往往可立竿见影。

这些似乎"小题大做"的观点手法，在当今也许会受至微词和贬鄙，但现在草率无力或渲染失真的设计不乏其数，不能不引起我们的反思。其实，许多传统的构图原则、处理技巧作为基本的判断准则和构成手段仍然有效，决不应一俱推翻。再者，我们现在接受的大凡是彻头彻尾的现代主义教育（譬如大学一年级皆以"构成"系列展开教学），却未加明晰其来龙去脉，把学院派视如瘟疫；而后现代派一开始就只是提供了一种姿态，许多人趋之若鹜，结果学步邯郸。

变通

变而求通，通达则融，融而生新。柳士英不拘泥盲从法式，敢于跳出法度之外，其建筑有浪漫诙谐的一面，许多地方即使以现代眼光审视，仍觉新奇：圆柱、半圆柱、退叠的竖线脚、弧线、曲拱状踏步，依稀有"后现代"气息。柳士英本人决无先知什么"后现代"，但"变通"是设计之大法，适当变形、夸张是极其自然的，正所谓盈满而溢，瓜熟蒂落。

涵忍

柳士英有朴素的建筑观，学术上不壮语惊人，设计不锋芒毕露，而重洗练凝重，有涵忍的胸襟和修养。如原湖大女生宿舍，外观虽实朴无华，但进门正墙仅作一圆窗，一枝梅花探进，清隽高逸。

回顾柳士英一生，"华海""苏工建筑科""中南土木建筑学院"，几多初创，足当在中国建筑发展史上留下重要的一页。然而，他却未获得应有的荣誉和知名度，甚至至今仍被疏漏于中国近代建筑史的研究中，究其原因：首先，华海及苏工建筑科存在时间不长，许多事已鲜为人知；第二，他早年来湘，扎根内地传播教育，远离上海等大城市，而这时正是大多国人建筑师在大城市广留作品的鼎盛时期。加之他本人对名利薄视，生性淡泊。其辛恳一生，更加值得追忆。

参考文献：柳道平.建筑学家、建筑教育家柳士英先生生平材料.湖南建筑.创刊号.

（本文节选自：魏春雨.纪念柳士英.建筑师（40））

二院

The Teaching and Research Centre No.2

原功能：教学楼，阅览室　　Original usages: teaching rooms, lecture theatre

建筑师：刘敦桢　　Architect: Mr. Liu Dunzhen

建筑年代：1925 年　　Construction end: 1925

建筑面积：2929.6 m²　　Area: 2929.6 m²

占地面积：1464.8 m²　　Site area: 1464.8 m²

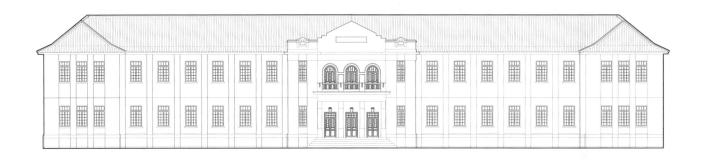

建筑概述

湖南大学二院，目前用作物理实验室，是刘敦桢先生少数实现的较大规模的建筑作品之一，也是刘先生早期建筑设计作品。自刘先生赴北京筹建营造学社后，他更多的是从事理论研究，因此这栋建筑的价值就不仅在于其精彩的设计语汇，同时还有了更多的历史价值和学术情怀。二院原功能是教学楼，当时一院即岳麓书院的斋舍（教学空间），"山"字形平面布局，西洋坡屋顶，门窗比例推敲严谨，尤其门头经过精心推敲，以柱廊、线脚、拱门及三角形墙体塑造了一个比例和谐、细节精巧的立面。主立面以砖砌筑的结构柱转角抹圆处理，结合当时质量优良的砖与精湛的做工，值得细细品味。1988年改造后，将原山形平面凹入的空间纳入室内一体。

· 门头细部
Details of porch design and architrave blocks

· 屋檐细部
Detail of eaves

· 浅弧形立柱细节
Details of piers design

· 柱头细部
Details of column capital

主入口立面 Elevation of main entrance

Building description

The Teaching and Research Centre Two is one of a few completed architectural work done by Mr. Liu Dunzhen, who was an early renowned architect in China, and it was a result of work at the early stage of his career. Mr. Liu Dunzhen had been engaging the theoretical studies immensely, and hence only a limited number of buildings have been left to us. This building is of historical and academic value owing to its uniqueness in design. Originally the plan of the building was in the epsilon shape but because of the building transformation in 1988 the two pocket spaces had been enclosed by two leaves of additional external walls, and become internal room. Moreover, the prefabricated bricks had been applied in this building construction completed with exquisite brickwork.

原图书馆
The Grand Library

原功能：图书馆

建筑师：蔡泽奉

设计时间：1929 年

竣工时间：1933 年

占地面积：1026 m²

Official name: The Grand Library, Hunan University

Architect: Mr. Cai Zefeng

Design start: 1929

Construction end: 1933

Site area: 1026 m²

建筑概述

由湖南大学教授蔡泽奉先生设计的国立湖南大学图书馆，当时是长江以南最大的图书馆，于 1933 年竣工，面积 1026m²。主入口柱廊采用的是花岗石制古罗马爱奥尼柱式，中央穹顶，欧洲文艺复兴建筑风格，顶部有八角塔，可作为观象台。1938 年被日本飞机炸毁，现仅存少量石柱，湖南大学如今位于牌楼口的校门即采用两根图书馆石柱为背景。

Building description

The Grand Library of Hunan University was completed in September 1933 with a total floor area of 1026 m². This library was the largest library at the south region of China, having an octagonal observatory built at the rooftop. It was built in the typical style of Renaissance having granite columns in Ionic order, central dome, etc. In 1938, it unfortunately became ruins because of Japanese bombing raids.

· 图书馆照片
Photo of the library

· 图书馆立面图纸
Design drawing of the elevation

湖南大學會書館建築設計會

比例尺八分之一吋等於一呎

湖南大學會書館

正面圖

側面圖

二舍
The Student Hall No.2

浅弧形墙面细部　*Details of piers*

原功能：二舍

建筑师：柳士英

设计时间：1935 年

Original usages: student dormitory

Architect: Mr. Liu Shiying

Design start: 1935

·二舍主入口
Main entrance

正门
Main entrance

· 楼梯1 · 楼梯2 · 门
Stair case1 · Stair case2 · The door

建筑概述

湖南大学第二宿舍受维也纳分离派风格的影响,具备现代主义建筑的诸多特征。入口处左右采用半圆形粗壮的柱墩,其上挑檐,柱墩与挑檐又并不重合。墙面则采用横向长水平线条绕窗台上下流动,收于阴角或绕成圆窗结束,整个立面浑然一体,具备柳士英先生建筑的特色,即强调整体的静止稳定与局部的变化流动,正门上方与侧面的圆窗与弧线更是柳先生常用建筑语汇母题,被人们称为"柳式圆圈"。21世纪初,湖南大学新建逸夫楼,拆除了二舍后部的部分宿舍,保留了最经典的东侧建筑和主立面。

Building description

This building is an example of Modernism of the time, and the design of the student hall was heavily influenced by Vienna Secessionism. The front porch is supported by two semi-columns with an overhang designed in the similar way. Horizontal flowing lines on the main facade of the building end at round windows.

· 二舍主入口上方
Above the main entrance

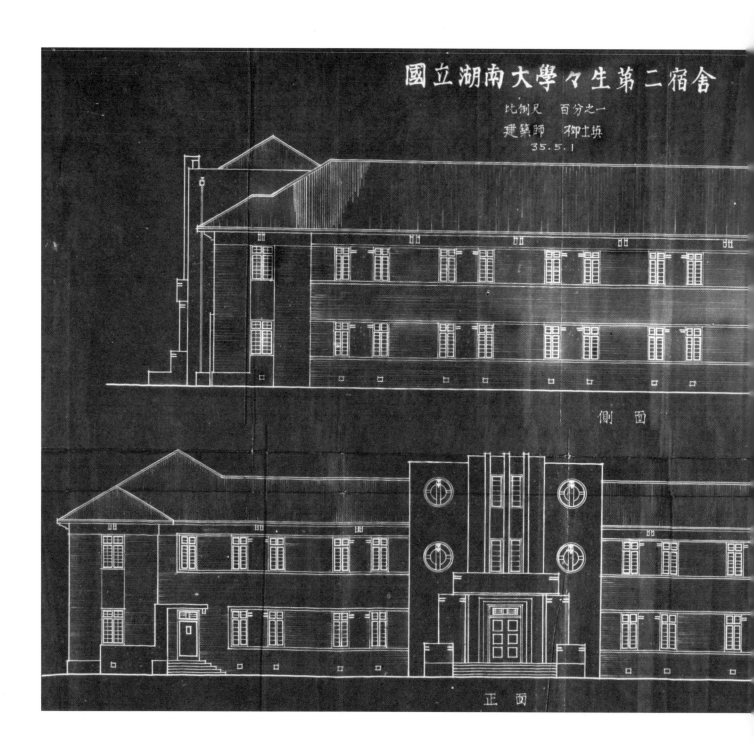

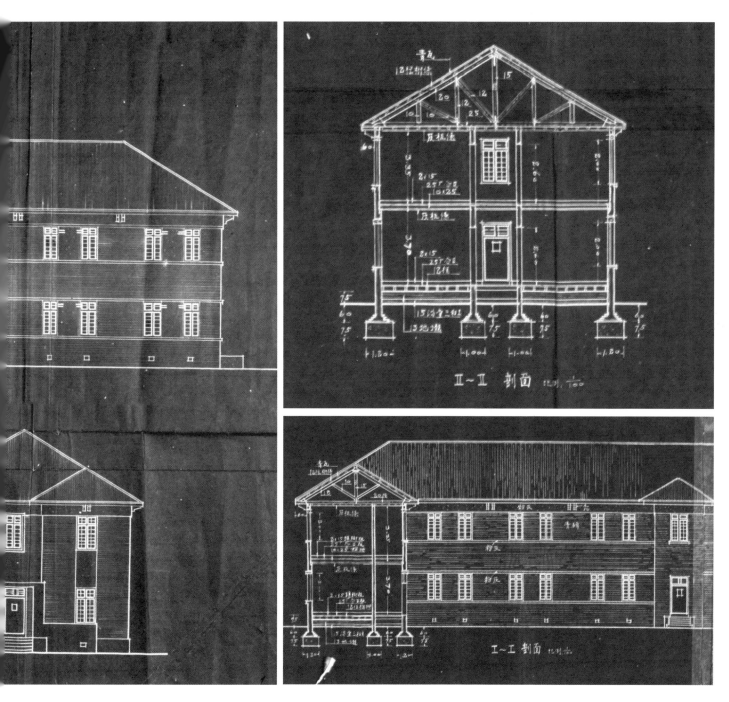

·设计图纸 - 立面图、剖面图
Working design drawing

科学馆
The Building of Science and Technology

原功能：科学馆	Original usages: office, museum, teaching lab
建筑师：蔡泽奉 柳士英	Architect: Mr. Cai Zefeng, Mr. Liu Shiying
设计时间：1933 年	Design start: 1933
竣工时间：1937 年	Construction end: 1937
加建时间：1948 年	Building extension start: 1948
建筑面积：4353 m^2	Area: 4353 m^2
占地面积：1500 m^2	Site area: 1500 m^2

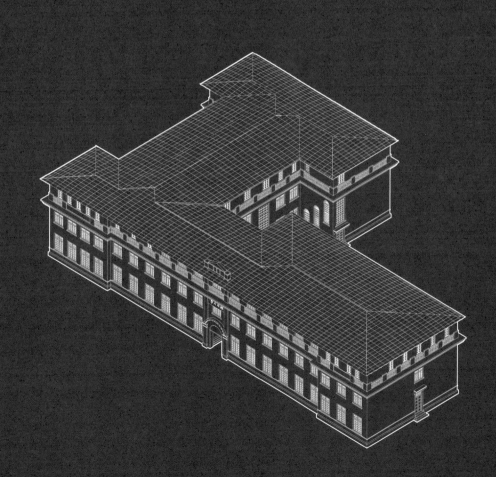

·科技楼主入口
Main entrance

建筑概述

由湖南大学教授蔡泽奉先生设计的科学馆，二层建筑，北侧主入口局部三层，1933年开始兴建，1935年建成。建筑为砖混结构，整个形体比例完美，体量均衡，风格采用了西方古典复兴样式，在大门、主入口门廊、阳台、檐口处均由细腻的线脚、转折线条、密檐凸块、圆形拱券处理，细节众多而又不显繁琐，显示了蔡泽奉高超的整体把握能力。1948年柳士英先生主持加建了一层，并未拆除原有的屋顶女儿墙和檐口，有的成为三层的装饰细节，有的以阳台处理栏杆，将加建部分巧妙地与原有部分衔接，屋面部分加琉璃瓦的坡屋顶，很好延续了原有建筑的风格，两位先生共同完成的这个作品，成就了湖大校园建筑中的又一杰作。

Building description

This building was initially two storey designed by Prof. Cai Zenfeng from the Department of Civil Engineering and completed in 1935 at first. It is brick-reinforced concrete structure built in the classic western style as the building details, for instance, cornice, archery door and so on, were typical style of classic renaissance. The scale of building is in a perfect proportion and its appropriate space volume helps to create delicate light shades. In 1904, an extension had been made by Mr. Liu Shiying to add one more storey on top of the old building. The original towers, parapets and cornice had remained but slope glazed tile roof of western style was built on the original flat roof. The results of the alternation had been another piece of significant works done by Mr. Liu

· 门细部 1

Details of entrance 1

· 门细部 2
Details of entrance 2

·屋檐细部
Detail of eaves

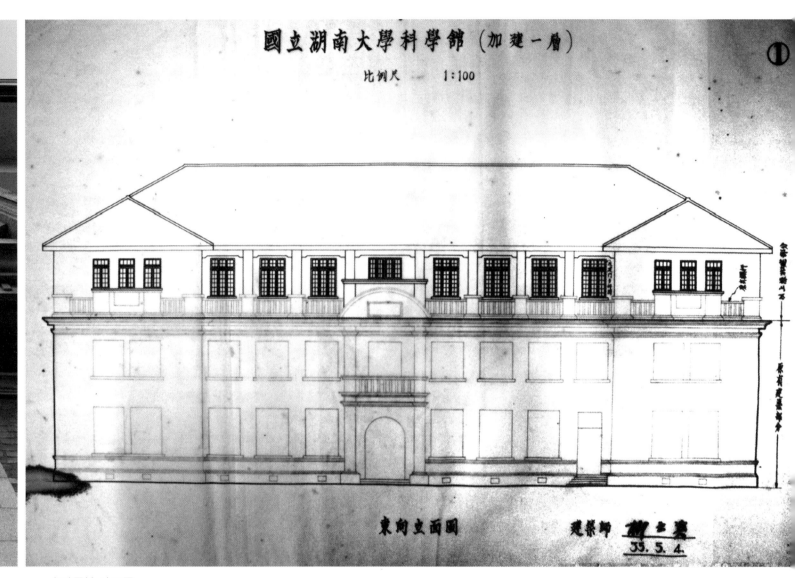

·加建图纸 - 立面图

Design drawings of the extension

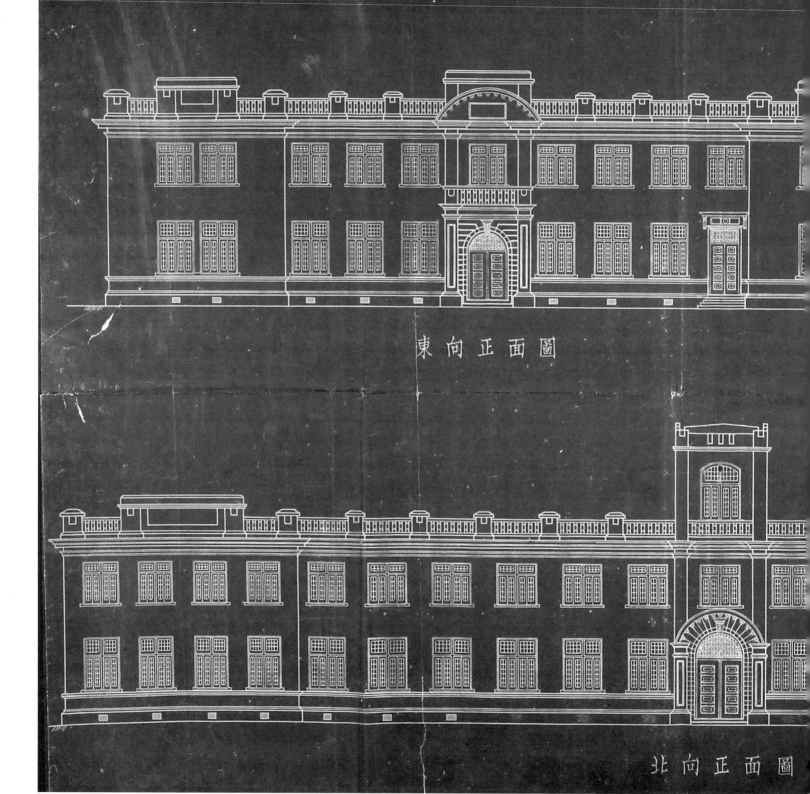

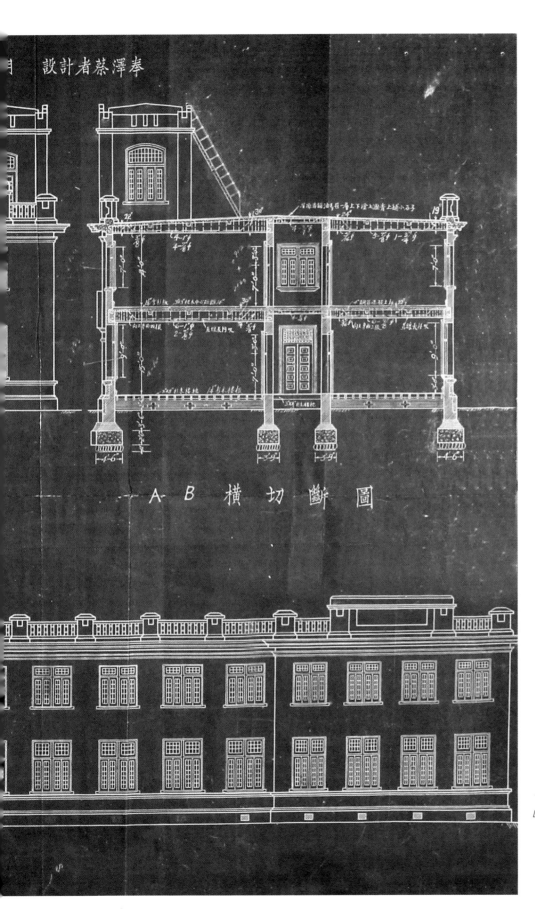

·设计图纸 - 立面图
Design drawings

九舍
The Student Hall No.9

原功能：第九学生宿舍	Original usages: dormitory
建筑师：柳士英	Architect: Mr. Liu Shiying
设计时间：1946 年	Design start: 1946
竣工时间：1946 年	Construction end: 1946
建筑面积：2127.8 m²	Area: 2127.8 m²
占地面积：1063.5 m²	Site area: 1063.5 m²

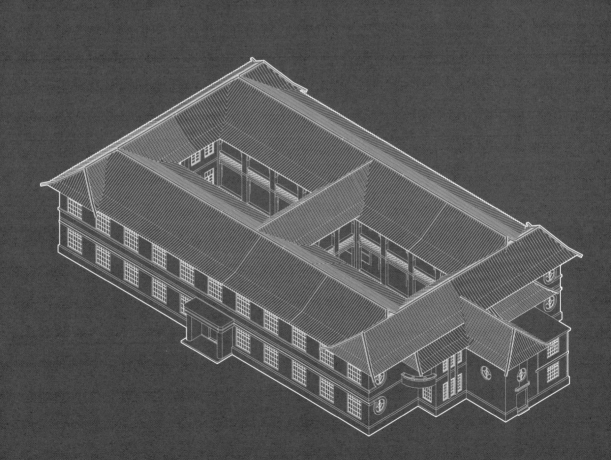

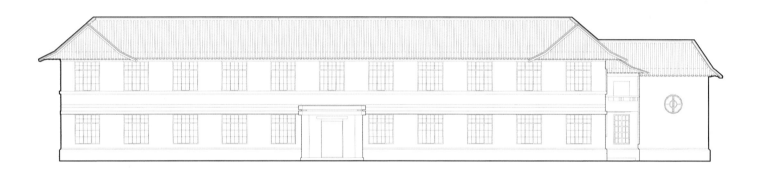

· 九舍主入口
Main entrance

· 窗细部
Details of window

· 楼梯
Stair case

· 拱形走廊
Arched corridor

建筑概述

湖南大学第九宿舍位于校园北部,电气院大楼西北侧,是一栋青砖材料的两层建筑,色调清新朴素。这一时期,柳士英先生在湖大设计了一系列教工或学生宿舍,空间与风格也类似。大体呈日字形平面布局,宿舍间以回廊连接,出现两个内部庭院,对外隔离而对内开放共享。九舍的特色在于内部回廊均为两层高的拱券结构,通畅明亮的走廊又有拱券的韵律,行走其间,别有风味。立面采用横向线条水平拉通带来建筑立面的动感,至转角处收于圆窗,仍是"柳式圆圈"的手法,但圆窗在不同部位做法又不尽相同,整体简洁而不乏细节,硬朗而不拖沓。

Building description

The Student Hall Nine is a dormitory for female students, next to the Student Hall 15 at the west. The building style is more or less similar to other buildings designed by Mr. Liu roughly at the same time. Central courtyards enclosed by the two-storey dormitory. Internal space is open to all dormitory rooms resulting in a good level of nature light indoor. The internal corridors are supported by elaborated arch-type structure, implying some cultural connotations. Furthermore the external wall had kept being Shimizu bricks coupled with grey black tiles for roofing and decorated with the 'signature' of Mr. Liu: fenestra rotundas and flowing lines ("Liu's Circle") . In short, the Student Hall Nine is a vivid demonstration of Mr. Liu's design approaches on building detailing.

· 柳氏圆圈
Liu's circle

七舍
The Student Hall No.7

原功能：七舍

建筑师：柳士英

建筑年代：1951 年

建筑面积：2511 m²

占地面积：979.7 m²

Original usages: student dormitory

Architect: Mr. Liu Shiying

Construction end: 1951

Area: 2511 m²

Site area: 979.7 m²

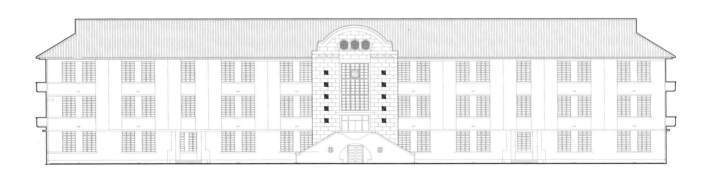

· 屋顶细部　Details of building elevation design

建筑概述

　　湖南大学第七宿舍位于湖南大学主轴线上，是中华人民共和国成立后修建的一栋宿舍。因宿舍位于凤凰山脚下，并未像其他几栋宿舍一样采用合院形式，而是将设计的重点放在了临主轴线的南立面上。南立面正中高耸的牌楼是其标志性建筑语汇，主入口采用半圆拱形入口，人字形的门头使得人可以从两侧进入二层歇台，有类似双首层的设计意向。通高贯通的竖向长窗使得楼梯间内的采光极佳，这在当时应属较为超前的设计。顶部以圆弧形高墙封口。在这个牌楼设计中浮雕语汇反复出现，底部拱门、上方曲面浮雕、并列的圆窗、圆弧形雨棚、"7"字的八角仿圆标牌、入口两侧的八角仿圆窗都为这个线条硬朗的立面增添了艺术气息。细看平面图纸，南侧窗间墙采用了三角形截面，并拉通上下，打断横向线条，因此形成了极其微妙的立面变化，尤其在阳光照射下会形成有趣味的阴影关系。整个建筑凸显了分离派的某些特点。

Building description

The distinguishing features of the Student Hall Seven are the curvature motifs which appear throughout this building, such as the three fenestra rotundas on the top middle of the southern elevation, an archery way on its main entrance, and the elegant curve reliefs on the wall. The varied designs of octagon adorn the south side of the building. The piers between the windows were devised in a triangular section so that a delicate light-shade effect could be created. The effect has then been enhanced by the external cement plaster on the breast under the windows in contrast to Shimizu bricks on the otherwise wall area. In all, the elevation design of the Student Hall Seven expresses a sense of Expressionism in colour application while Secessionism in the building detailing.

· 细部 1 · 细部 2
Details 1 Details 2

· 细部 3 · 细部 4
Details 3 Details 4

湖南大学第七宿舍

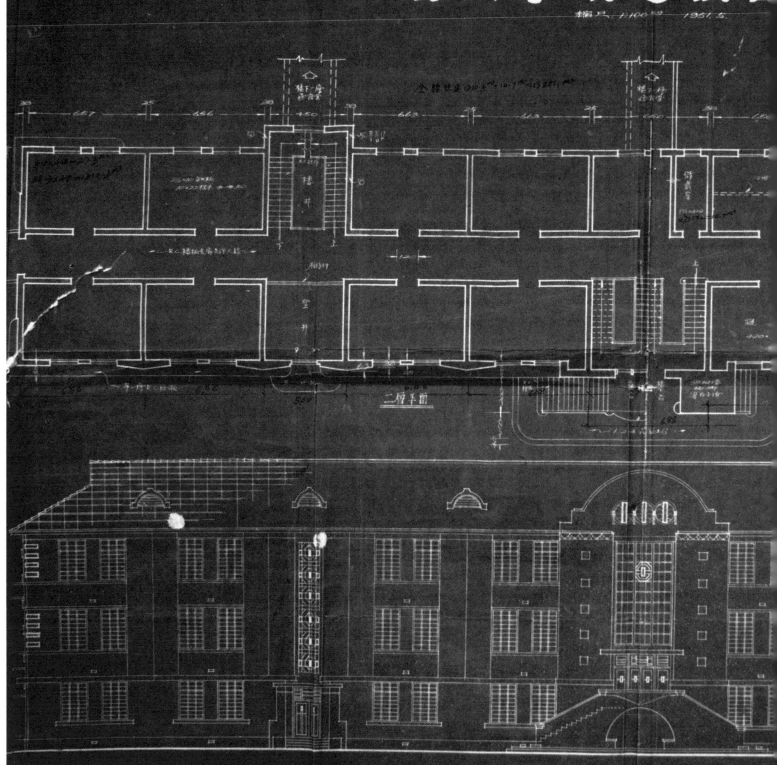

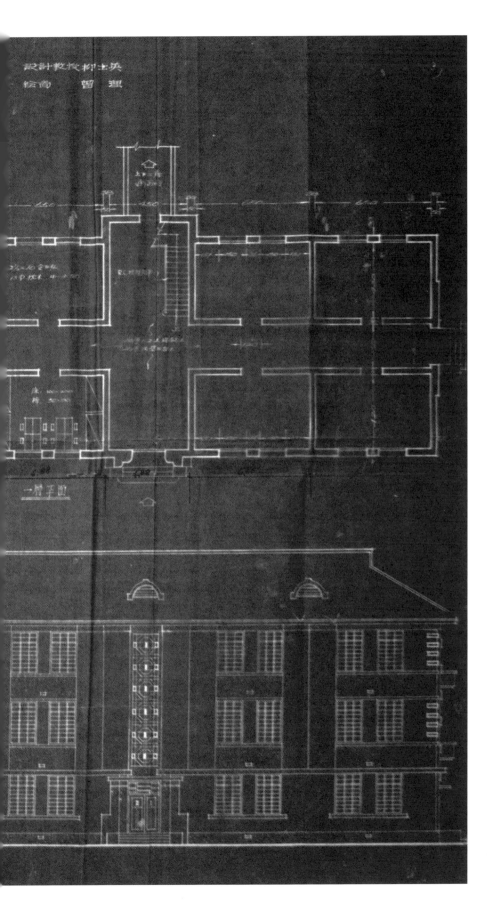

· 设计图纸 - 平面图、立面图
Design drawings

胜利斋
The Shengli Residence Hall

原功能：胜利斋教工宿舍

建筑师：柳士英

设计年代：1950 年

建筑面积：2211 m²

占地面积：1321 m²

Original usages: residence hall for staff of the university

Architect: Mr. Liu Shiying

Design start: 1950

Area: 2211 m²

Site area: 1321 m²

· 胜利斋主入口
Main entrance

· 庭院
courtyard

· 东立面细节
Details of east elevation

建筑概述

胜利斋位于岳麓书院东南侧，是教职工宿舍，层数两层。依据地形条件，建筑平面采用了十字形布局，单侧外廊使得各个部分的宿舍都能面对四个庭院，由于庭院有两进形成了纵深感，人们在宿舍内走动时似有步移景异的效果。西侧紧邻岳麓山，中部设计了小食堂，功能分区明确，也避免了出现西晒的宿舍。一层圆窗使得建筑有动感与活力，也是"柳式圆圈"的再次应用。胜利斋的手法和布局及细节都已经是非常明晰的现代主义建筑，是柳士英先生较为成熟的作品。因偏于一隅，许多人都忽视了这个低调平朴的建筑，其空间变化的精妙亦需要静心体会才能感知。

Building description

The Victory Residence Hall is one of representative works of Mr. Liu Shiying. Overall layout was organized perfectly according to the space usage and geographical typology surrounding the building. Correspondingly its plan became asymmetric and a canteen was built in a relative isolated place at the west wing of the residence hall. Main elevation was designed by using the typical expressionist measures of Mr. Liu, such as "Liu's Circle" treatment. A clever transition between internal corridors and external ones had been made at the north side in an attempt to enrich constructive elements on the main elevation. In all, the hall could be the masterpiece of modernist architecture designed by Mr. Liu in his time.

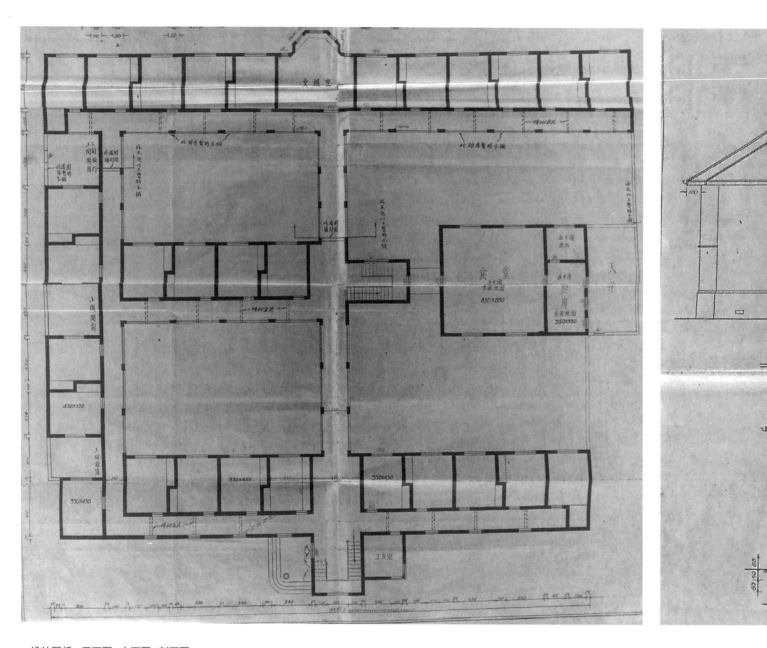

· 设计图纸 - 平面图、立面图、剖面图

Design drawings

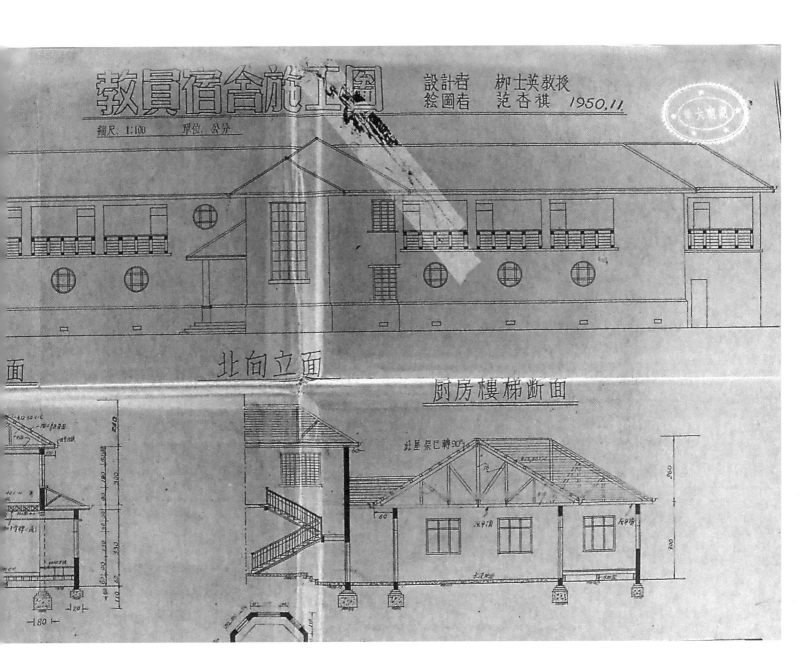

原六舍
The Former Student Hall No.6

原功能：湖南大学六舍　　Original usages: dormitory

建筑师：柳士英　　Architect: Mr. Liu Shiying

建筑年代：1950 年　　Construction end: 1950

建筑面积：3500 m²　　Area: 3500 m²

占地面积：1216 m²　　Site area: 1216 m²

·主入口
Main entrace

· 大门 Main entrace

建筑概述

　　抗战时期湖南大学迁往湘西，校舍遭到日军轰炸、焚烧，损失惨重。柳士英主持设计了一批新的宿舍建筑，都是采用木屋架、木楼板、青砖、青瓦、清水墙面，清一色的地方材料，因地制宜，在当时那样困难的条件下，教工学生宿舍能建成这样水平，凝结了柳先生不少心血。原六舍就属于这一时期的作品，建设时值抗日战争胜利不久。六舍为砖木结构，也设有庭院，而且平面与九舍极为相似，均为单面外廊组合四合院，建筑内部形成日字形回廊，用红砖砌筑花样作为走廊的栏杆。但较之九舍更加朴实，简洁，少了很多细部，如九舍的券廊和门上的细部等。建筑立面造型简洁，重点处理入口部位，入口朴实大方，转折进入豁然开朗的庭院；入口平面呈弧形环抱之势，设立半圆柱以及弧形的入口雨棚。

Building description

This student hall was built just after the war of resistance against Japan. During the World War II, Hunan University had moved to remote area at the west side of Hunan province. Most buildings, accommodation in the campus had become ruins as results of bombardment. The student hall was one of the student accommodation design projects managed by Prof. Liu Shiying. The construction methods adopted according to the local conditions of the depressing time were to use timber frame structure, wooden flooring, and some local building resources, such as black grey bricks, grey tiles, Shimizu bricks, etc. The results of these construction works turned out to be satisfactory owing to the great input of Mr. Liu. The layout arrangement of the Student Hall One is similar to that of the Student Hall Nine (which is given an account later.). There are single-sided corridors, two courtyards in the centre, balustrade of red brick work, and simplistic elevation design adorned with "Liu's Circle" design. Much attention of the design had been given to its entrance, which was flunked by two semi-columns and covered by a canopy with a curved outline.

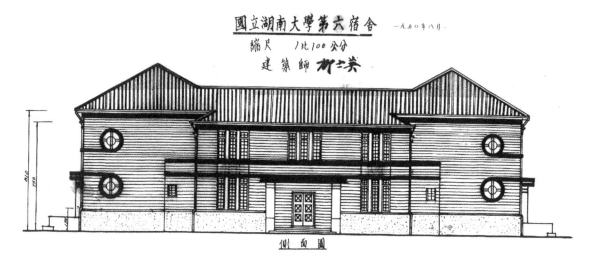
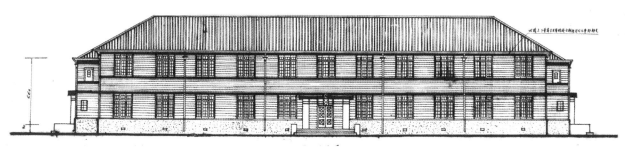
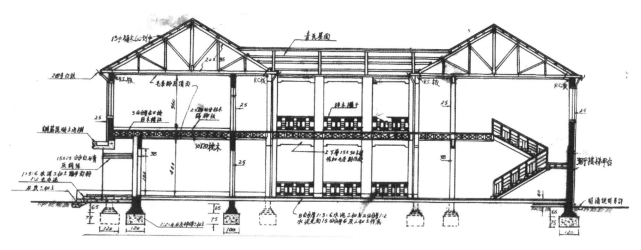

· 设计图纸 - 立面图、剖面图

Design drawings

工程馆
Building for the Department of Engineering

原功能：工程馆　　　　Original usages: Building for the Department of Engineering

建筑师：柳士英　　　　Architect: Mr. Liu Shiying

设计时间：1947 年　　　Design start: 1947

竣工时间：1953 年　　　Construction end: 1953

建筑面积：7430 m²　　　Area: 7430 m²

占地面积：2060 m²　　　Site area: 2060 m²

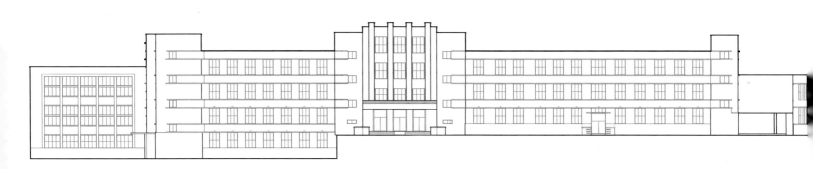

· 北面弧墙
Curved wall at the north side

·南立面全景　　*Perspective view of the south*

· 弧形细部
Details of the curvature design

· 弧形细部圆形墙面
Details of the curved wall design

· 波浪形墙面
The parenthetical wall

建筑概述

湖南大学工程馆是柳士英先生现代主义风格最典型的代表作。一方面，建筑为砌体建构，表面部分涂刷素混凝土；另一方面，未用任何细小装饰，把细小线条变成了面的延续。平屋顶高低错落。建筑楼梯间的圆弧形墙体，墙面上通长的水平线条，以及圆弧形窗檐，窗台和窗口墙体，都具有典型的德国表现主义流动线条的造型特征。而各种曲线面处理的承前启后，有起有落，体现了他将这些独立的元素通通揉进建筑肌体之中的设计方式。凡线条应有所交代，找到归宿之处，兼有"起、承、转、合、让"的节奏变换。工程馆背立面原本平整连续的墙柱做成三角形的轻微波浪形状，使墙柱形成一阴一阳的对比，从而突出了建筑的主体形象，增强立体感。立面正中垂直通窗体现了分离派的审美取向，即歌颂现代社会的速度美。长长的条窗与间隙突出的柱子形成了一种向上感，是高速生产力水平发展在建筑的意向投射。

· 垂直道窗与竖向线条

Details of vertical adorned lines and windows

Building description

The North Building could exemplify the design style of Prof. Liu in his career as a modern architect. Two leading schools of modern design, Vienna Secessionism and German Expressionism, influenced the thinking and aesthetics of architectural design of Prof. Liu when he was in his academic pursuit of architecture in Japan. On the one hand, the staircase is screened inside a curve wall, and windows were designed with round lintel, curved sills and breasts underneath. These building designs reflect the influence of German Expressionism featuring flowing lines, but are imprinted with his own style of "Dynamic in static" as a flowing line is completed with a circle, which is the treatment thereafter given a name "Liu's Circle". On the other hand, the design of vertical windows embodies the Secessionist style, celebrating the developing speed of modern society. The long windows spreading vertically coupled with pilasters in between provoke a sense of soaring-up-to-sky mirroring the scene of high productivity of the modern time.

· 楼梯细部
Details of stair case design

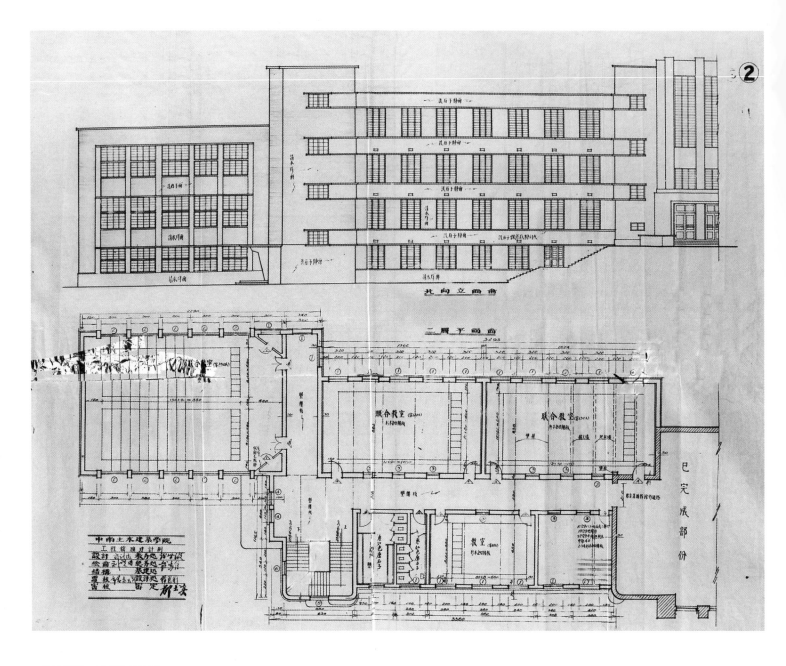

· 设计图纸 - 立面图、平面图

Design drawings

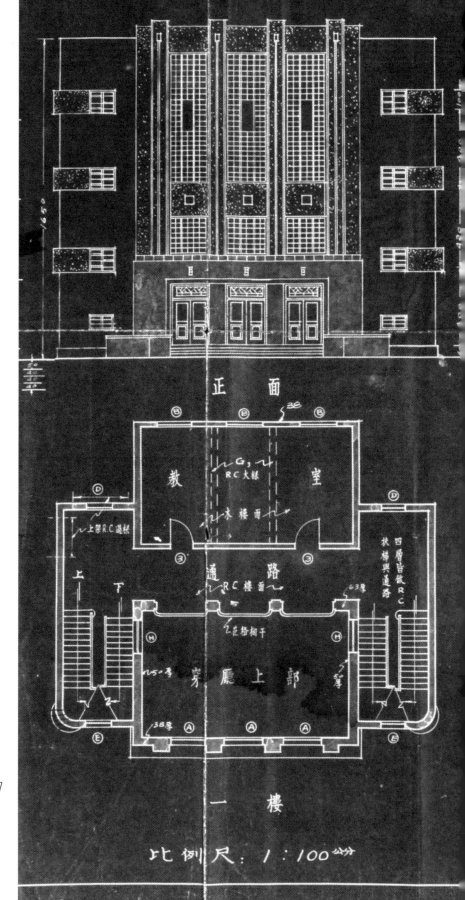

· 设计图纸 - 立面图、平面图
Design drawings

大礼堂
The Grand Hall of Hunan University

原功能：大礼堂

建筑师：柳士英

设计时间：1951 年

竣工时间：1953 年

建筑面积：2142 m²

占地面积：1094 m²

Original usages: auditorium, assembly hall

Architect: Mr. Liu Shiying

Design start: 1951

Construction end: 1953

Area: 2142 m²

Site area: 1094 m²

主入口
Main entrance

·立面细部
Details of elevaton

建筑概述

大礼堂在经济困难条件下仓促上马，为了求得以最经济的办法，最快的速度和最大容量建造，采取了大跨度木屋架、最经济的钢筋混凝土构件断面、普通水泥粉刷、尽量减少辅助面积等措施，仅费旧币25亿元（人民币25万），体现千年学府传统的庄重典雅造型。

Building description

The Grand Hall was built in times of economic stress. The challenges had been tackled by the outstanding architects of the time, who had used large-span timber structures combining structural components of reinforced concrete and applied cement plaster in a large scale. Thanks to the short construction time, limited ancillary building area, and economic construction methods, the over cost of the hall was about 2.5 billion old Yuan (equivalent to RMB 250k).

· 转角斗栱
Corner corbel-bracket set

　　大礼堂以圆形装饰为题，舞台原有圆形框饰以增强人流动感，而两球之间的花饰，和舞台上部三组圆圈的边饰，外圆窗的花饰线条等又起到稳定作用，使之"动中有静"。波形、折形墙面的处理，诸多刚柔动静的表现手法，都独具匠心。

　　大礼堂采用官式绿色琉璃大屋顶，但又不拘泥于法式做法，力求淡雅、明快，礼堂装修采用国漆红黑色主调，点缀金饰，既存湖南楚汉文物的艺术特色，又颇具现代意味。外墙粉刷以普通水泥掺和黄泥石灰碎玻璃等，达到了经久耐用的目的和较好的彩色效果。平凡显新意，清淡求深情，正是其创作的独到之处。

· 背立面　*The elevation design*

    The interior decoration is a mix of modern design and traditional Chinese fine art of Xiang, which dated back to Han dynasty 2000 years ago. Chinese lacquer paint is dominant with a garnish of golden design. A mixture of cement, yellow limestone and glass cullet had been used on the exterior wall, and hereby a long last colour shade of external wall had been achieved. The overall design of building turns out to be simple but elegant. Prof. Liu Shiying recalled that although the nationalist style of architectural design was popular in China at the time, he as a modern architect still insisted on the implementation of modern designs in this particular building, notably the detailing design. Therefore it was hard to tell which style category this building design could be identified to. In general, this building was the result of the first design attempt to create a unique modern Grand Hall in this university campus. From the view-point of prof. Liu, it was an innovative architectural work but still needed further work-input to refine its overall design.

柳士英回忆说:"当时人们主张采用自己的民族形式,我仅仅在轮廓上着眼,在细部上是有自己的手法的。天幔、楼梯、门窗以及台口装饰都是有显著的不同风格,很难说是中国或是西方的样式。大礼堂是湖南大学第一个新型建筑物,也是我第一次的尝试作品,设计意图是革新的,式样是不够熟练的,不入于洋,不入于墨,有我的个性,亦中亦外,亦古亦今,是我的全貌,只因为实践仅限于尝试,素养还不够充实,从整体看,很难说是一个完整的作品。"

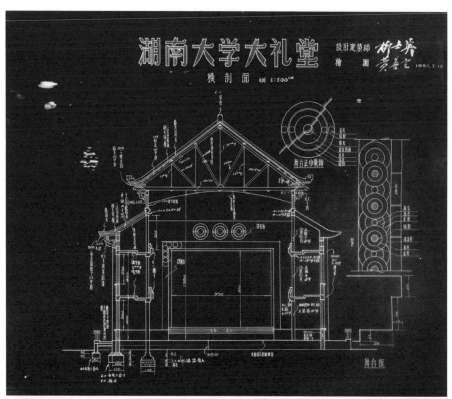

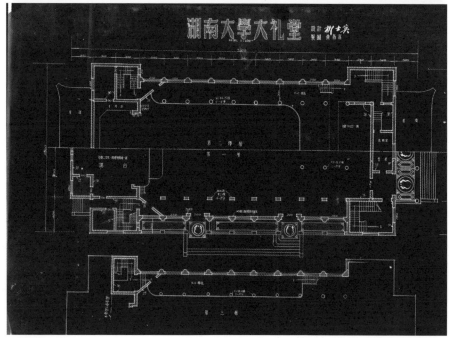

·设计图纸 - 剖面图、平面图

Design drawings

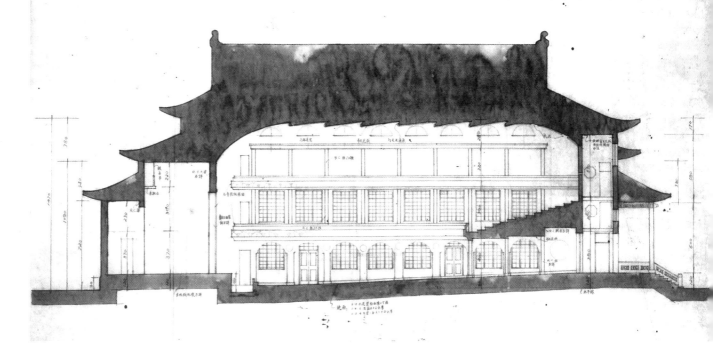

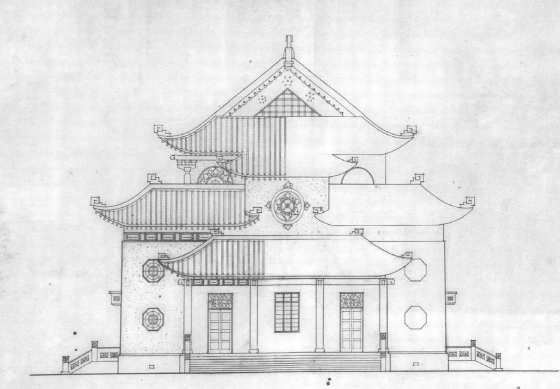

· 设计图纸 -
剖面图、立面图
Design drawings

大南湖

側面圖

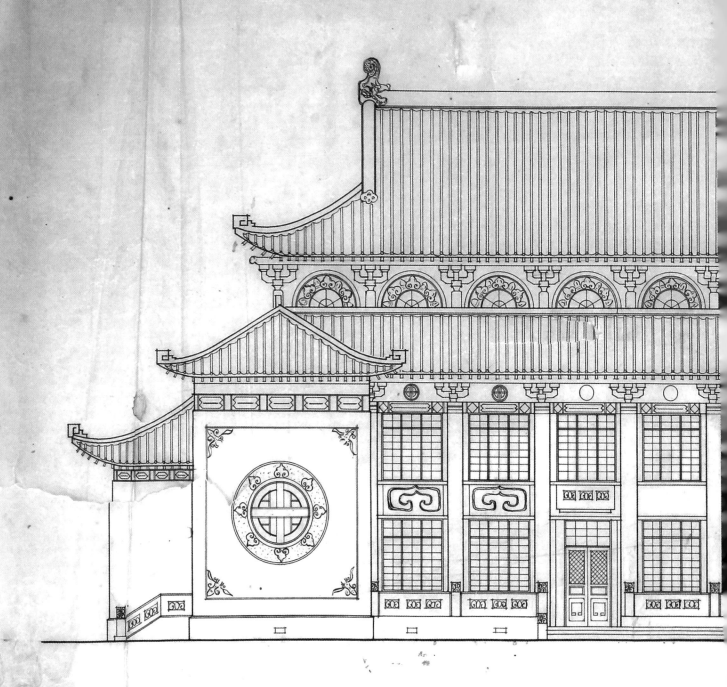

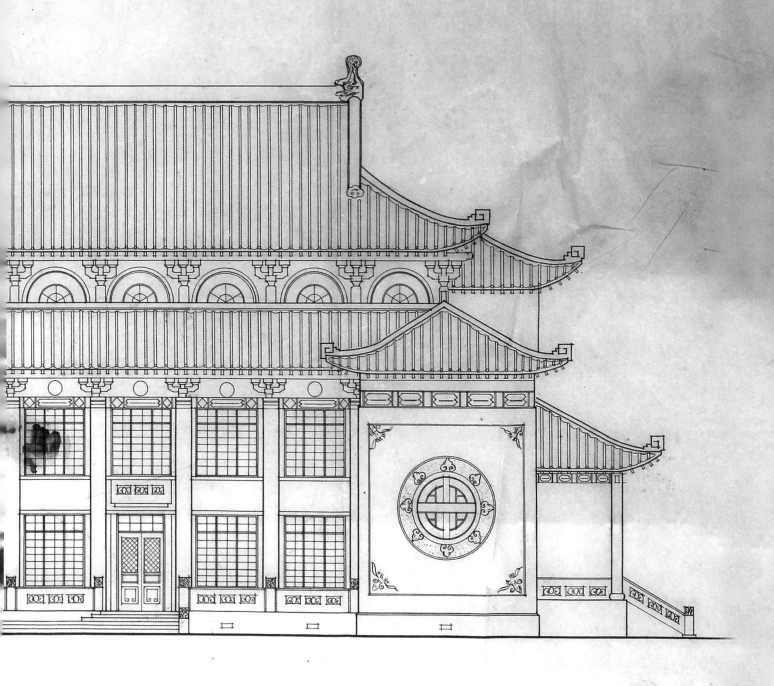

大禮堂

設計教授 柳士英
繪圖 黃善言

例 1:100 cm.

图书馆
The Library

原功能：图书馆

建筑师：柳士英

建筑年代：1946 年开工修复遗址，1948 年竣工，后 1951 年扩建，1954 年竣工完成

建筑面积：2362 m²

占地面积：1097 m²

Original usages: library

Architect: Mr. Liu Shiying

Design start: 1946

Construction end: 1948

Extension start: 1951

Construction end: 1954

Area: 2362 m²

Site area: 1097 m²

·图书馆鸟瞰图　　Bird view of the library

建筑概述

图书馆所处位置较为特别。沿岳麓山清风峡而下，经岳麓书院，这里是被读书人称为"风水宝地"的地方，形成了湖南大学建筑的第一根中轴线。在这根中轴线上，建有三座图书馆：岳麓书院御书楼、老图书馆和现在的图书馆。由建筑大师柳士英设计建造的老图书馆，1948年竣工，有书库三层，目录厅和办公房若干。碧瓦红砖，飞檐画栋，形式古朴。1947年馆藏图书79182册。该馆曾于1951年扩建，总面积达2300多平方米。她与岳麓书院前的大礼堂组成了极富民族特色的仿古建筑群，成为湖大校园建筑的"经典"。

Building description

This old library is located on the central axis of the university campus, like other two libraries in the campus. The old library was designed by Prof. Liu Shiying and completed in 1948. It had three-storey storage rooms for books, one category and load service hall, as well as a few offices. In 1947, the book storage was up to 79182 copies and a building extension had been made in 1951 so that the overall floor area had reached 2300 m^2.

· 竖向通窗　The vertical long window

· 窗细节　The details of windows

　　该建筑为砌体结构，中国传统大屋顶式样，采用官式绿色琉璃瓦，色彩淡雅，明快；建筑造型中加入西方早期现代主义手法，特别是正立面上通贯多层的竖向长窗，是典型的维也纳分离派的造型特征。檐下、墙壁等细部装饰部分又具有浓郁的中国风格，使之成为东西合璧的优秀建筑。建筑两翼以洞门小井联系，适应长沙气候条件，以利分流。整个建筑实用典雅，与大礼堂组成的建筑群为《中国现代建筑史》所载入的湖南四处建筑之一，也是近现代保护建筑。

· 拱楼　*The details of stair case*

  The old library is a classic example of campus buildings in the Hunan University. Its Chinese traditional roof-top design employs green glazed roof tiles and reflects particular Chinese aesthetics of architecture. The ornament details under the eaves and wall showing rich Chinese culture. A certain western style of modernism has meanwhile been apparent in the building details, such as the vertical windows at the main facade characterized as Vienna secessionist design. The design of the building in general is a mix of the western and Chinese architecture and follows some aesthetics of modernism, for instance, simplism, functionalism, etc. Besides the Grand Hall in the university campus, this building has been listed in the neoteric protect buildings of China and gone down in the History of Chinese Neoteric Architecture.

·东侧庭院
The east courtyard

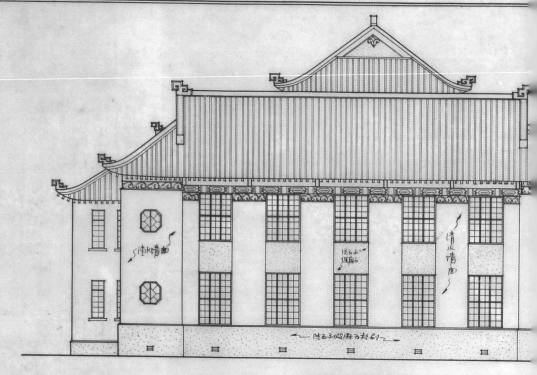

南向側面圖

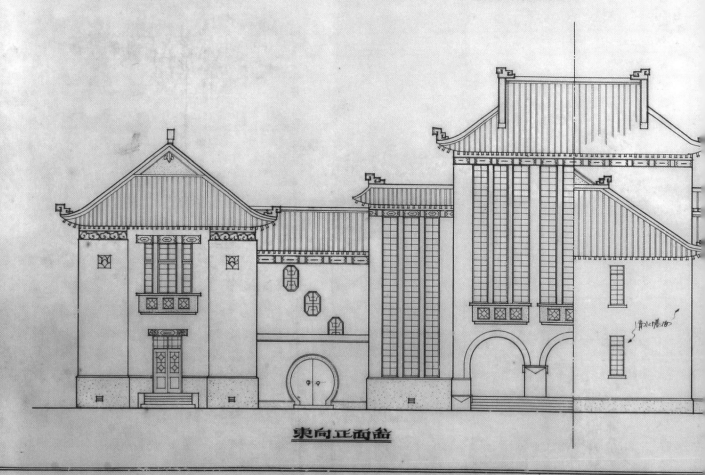

東向正面圖

湖大圖書館
擴充計劃
比例尺：百分之一
一九五二年五月

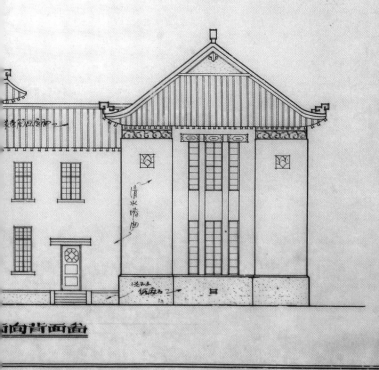

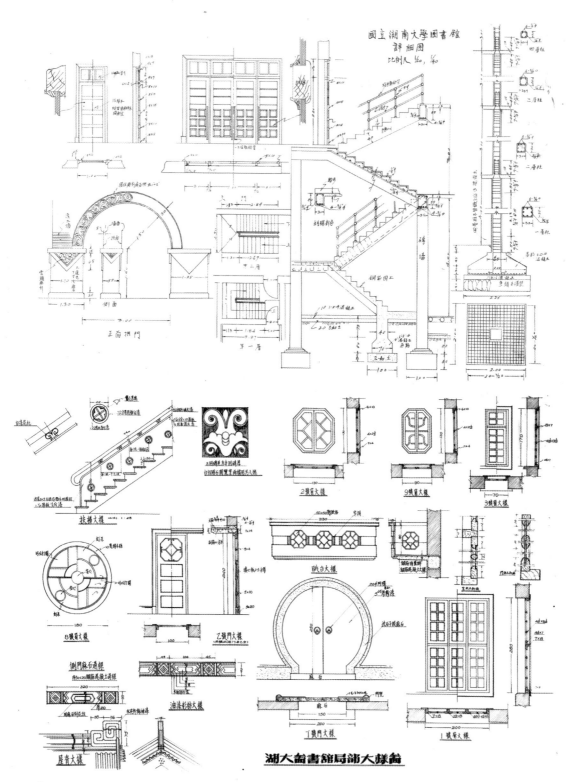

营造篇 新地域的建构
（现代湖南大学发展时期）1954 — 现在

The Campus Architecture New Regionalism and Re-construction (Hunan University):1954-Present

20 世纪 60 到 80 年代学校的建筑量并不大，这一时期建成的有化工楼、电气楼、中楼、东楼等一批较朴实的校园建筑。然而自 20 世纪 80 年代起，由于学校迅速的发展，校园面积不断扩大，故而在原有校园中心区的基础上进行了大幅的改造、扩建及新建。这是校园的第二次传承与变异，书院仍然是湖南大学的文化中心，校园的文脉与特色得以保留。学校原有中心区南移，在天马山西麓新征用地进行教学建设，将校园的发展引向南面。于是形成了相互垂直的两条轴线：一条为文脉轴，起始于岳麓书院，垂直于岳麓山，收于湘江畔牌楼口，所谓"藏之名山"；另一条为新的轴线，我们称之为教学轴，平行于岳麓山，新老校区通过轴线结合，该轴线在运动场处进行了扭转，形成一个完整而又富于变化的空间序列，所谓"纳于大麓"。

In 1960 -1980s, there were only a number of purpose-built educational buildings, such as Building of the Department of Chemistry, Building of the Department of Electrical engineering, Central Building for Education, East Building for Education, etc. They were built in the architectural style of modernism emphasizing on simplism and functionalism. Since 1980s, the university campus expanded quickly and has experienced a large scale of redevelopment upon its historical central area. During the second period of transformation in the campus, the position of Yuelu Academy as a culture centre remains intact. Apart from the old central axis of planning as a reference to "hold historical campus of scenic beauty", a new axial line has developed in the campus that intersects the old one at a right angel. The new one parallel to the Yuelu Mountain range and suggests a new planning strategies for "creating a modern cradle land of great talents", notably for the educational development in the campus .

从总的校园发展历程来看，可以发现其中交融着几种关系，主导了校园空间形态的变迁。

一、校园与城的关系：书院选址与城市相对脱离，须经牌楼口摆渡登岸方可进入学府修身求学之地。而现代大学则在形态上与城市气息相吻合，较为现代。

二、校园与山的关系：岳麓书院最早兴起于宋朝佛僧兴办佛堂，并一直延续到元、明、清，它的主要建筑藏于山林峡谷。而现代校园规划中，岳麓山更多的是做为一个大的屏障，学校的建筑逐渐与山脱离，尽量不把活动导入山间。

三、新与旧的关系：早期岳麓书院是自然生长的发展模式，如岳麓书院的斋舍，顺应山势以及地形自然生长。而近20年的发展，校园内新建了一大批现代建筑，在建筑形体、材料及尺度上更多是顺应场地，而且没有完全克隆传统的材料与风格，在总体校园环境中形成了风格相对明显的区域，但因时因地的新地域手法使岳麓书院的文脉与空间形态得以延续。

In the overall course of the campus development, it is found that the dominant relations exerting considerable influence on the spatial transformation of the campus can be identified as below:

1) Spatial relation between the university campus and the city. With the changes over the years, the university campus has interwoven the city fabric intensively, and is not the same as it used to be lying in remote region to the urban area.

2) Positional relation between the campus and the Yuelu Mountain. The Yuelu Mountain nowadays functions as a natural protective screen overlooking the campus rather than a part of redevelopment site for campus architecture as it used to.

3) Historical relation between the campus architecture of the past and the present. The new campus architecture has created its own identity from the angle of regionalism, managed to integrate the different styles of architecture in vary construction approaches which are responsive to the surroundings and the times, and kept striving to achieve sustainability of the campus development.

书院博物馆
The Museum of Chinese Academy

2004 年，在湖南大学校园岳麓书院旁建设了"中国书院博物馆"，并选址于原柳士英先生设计的临时校舍"静一斋"，与岳麓书院原有的讲学、藏书、祭祀三部分，共同形成了一个完整的书院建筑群，共同表达文人建筑的儒雅气质。

一檐青瓦，一片素墙，一簇翠竹，一通回廊，内向式的开放，简洁精炼，乃是中国书斋的精魂，它剥离了纷繁芜杂，为人们提供了一个去除杂念，去欲思静，精神归一的场所，整齐与干净本乃"斋"的原意。

Verdant tile, white wall, bamboo, winding corridor, introverted open, clean and refining, these are the spirit of Chinese studying room. They strip the messy and complexity, providing people a place without distracting thoughts and desire. Tidy and clean are the original intention of "Zhai".

设计：魏春雨 齐靖
设计时间：2003-2004 年
竣工时间：2014 年
规模：16291m²

Architect: Mr. Wei Chunyu Mr. Qi Jing
Design start: 2003-2004
Construction end: 2014
Area: 16291 m²

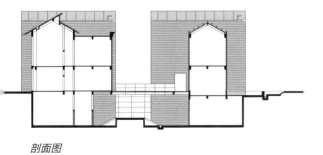

剖面图

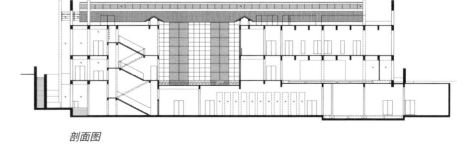

剖面图

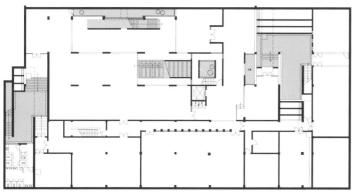

负一层平面图

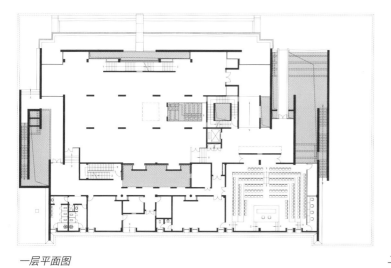

一层平面图

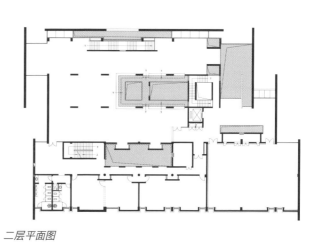

二层平面图

生物技术综合楼
Building of Bio-science

设计通过将承重结构体系与围护结构体系两者分离、拉开，产生了"缝"，光从多角度进入，反射，散射，折射，投影等多重关系，产生了奇妙的效果，这一切缘于"拉开"的缝隙。由于形体呈现片段状，前面有棵桂花树遮挡，直至现在，许多人仍未知它的存在，这个设计是边庭空间与复合界面构成的一次实践。

The first attempt to implement the concept of complex interface design on buildings could be the project of Bio-science building. This building has the facade system detached from the building weight-bearing structure, creating a gap in between. An intriguing light effect therefore has be produced in such an in-between space owing to light reflecting, refracting, casting, either/or scattering. The final results of this construction work have been concealed and cannot be found out easily by viewing from the outside.

设计：魏春雨
设计时间：2001 年
竣工时间：2002 年
规模：3280 m²

Architect: Mr. Wei Chunyu
Design start: 2001
Construction end: 2002
Area: 3280 m²

复临舍教学楼
Building of Fulin

复临舍教学楼通过公共中庭、宽敞的走廊、明亮的光环境体现着开放的大学教育理念。面对西面的岳麓山景观，通过大面积的玻璃幕墙取得对这一校园最大景观要素的引入。形体构成采用复合界面的手法，通过立面复合化处理，形成立体构成式的视觉效果、摇曳的光影效果，并通过主入口的弧形玻璃，建立了东方红广场和室内中庭间的轴线。

Building complex of Fulin emphasizes on the communication between the inside and outside by creating "grey space" at the entrance, through which a connection between the Dongfanghong Square and internal atrium has been established. The large curtain window gives a vivid view to the Yuelu Mountain at the west, displaying a gesture to welcome the outside environment into the inside. The design of facade has been dealt by utilizing the methods of complex-interface. Especially the final results of the work, such as the spacious periphery courtyards, airy open corridor arrangement and bright interior space coupled with a soft facade design are all providing a lively image of the open culture of the university education.

设计：魏春雨 邓毅
设计时间：1999 年
竣工时间：2000 年
规模：9550 m²

Architect: Mr. Wei Chunyu, Mr. Deng Yi
Design start: 1999
Construction end: 2000
Area: 9550 m²

法学院、建筑学院建筑群
The Building Complex for the School of Law and the School of Architecture

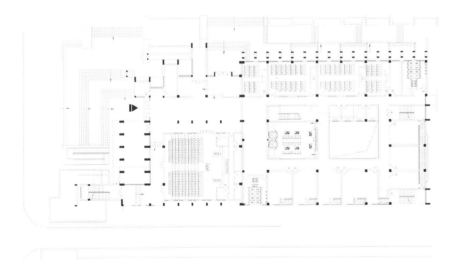 

　　法学院·建筑学院建筑群体现了"场域"意识，让"建筑从土里生长出来"。在自身限定的区域中，设计从场地到建筑统一采用水刷石饰面，对光形成漫反射，使建筑"折旧"而易于融入周遭。结合坡地地形设置的舒缓台阶踏步，在逐级上升过程中自然渐变为建筑本体，完成一个共同的"场域"。尝试用湘江中的石子作为原料，通过水刷石这种传统工艺做成外墙。把石子墙面延伸到室内，水刷石对光具有漫反射和折射效果，在不同季节、天气和阳光角度下可以呈现不同的色泽和明暗感觉。

 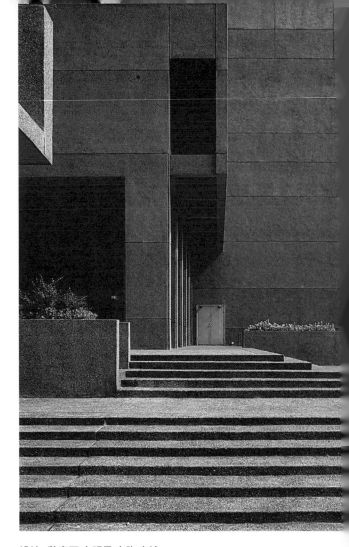

The building complex for the School of Law and School of Architecture seeks for the architectural succession of tradition and integration in the campus. The design principles of Regionalism have been displayed when designing this complex. Granitic plaster, a local sourcing of building materials from nearby Xiang River, has been applied in a large quantity on the interior/exterior walls of the building in order to create a material identity of the place. Terraces gently run down along the building's north and west wings. A building responsive to the site where it grows out of therefore takes into shape.

设计：魏春雨 宋明星 李煦 齐靖
设计时间：法学院 2002
　　　　　建筑系馆 2003
竣工时间：法学院 2003
　　　　　建筑系馆 2004
规模：法学院 10784m²
　　　建筑系馆 5000m²

Architect: Mr. Wei Chunyu Mr. Song Mingxing Mr. Li Xu Mr. Qi Jing
Design start: 2002-2003
Construction end: 2003-2004
Area: Law school 10784m²,
　　　Architecture college 5000m²

工商管理学院楼
Building of the Business School

　　传统聚落的"图—底"关系是非常生动的,并且往往是可以互换的。在工商管理学院设计中,保留了基地内的一棵树,使"底"有了精神场所与归宿。场地的原住民——一棵树,成为空间主角。围绕树,建筑以玻璃及开放空间展开,而其余部分则多以实体包围,强化与树的共享、交流氛围,营造了以树为中心的场所感。

The fascinating interaction of 'figure and ground' underlying the Gestalt aesthetics can be epitomized by the traditional settlement arrangement. The building of the Business School aims to adopt these traditional patterns of the regional buildings and reflect some traditional value. Its design concept was inspired by the attempt to conserve a camphor tree, standing in the center of the site, the 'ground'. The central tree becomes a major actor in the 'ground', while the 'ground' becomes a stage. The building layout is designed to be open toward the center and organized carefully around it. Meanwhile the solid walls facing outward from the central 'ground' have been constructed with purpose to create a concentric communal place, by which the leading role of the tree has been emphasized and the design results has achieved to some extend environment harmony on the site.

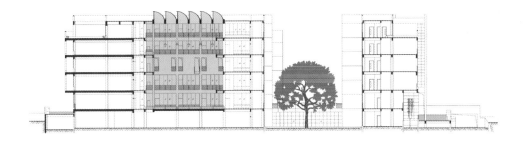

设计：魏春雨 宋明星 李煦 齐靖
设计时间：2003—2004 年
竣工时间：2006 年
规模：16291m²

Architect: Mr. Wei Chunyu Mr. Song Mingxing
Mr. Li Xu Mr. Qi Jing
Design start: 2003-2004
Construction end: 2006
Area: 16291 m²

软件学院楼
Building of the School of Computer Science

软件学院整体形态以合院形式沿四周有机组织成巨型的庭院空间，强化了合院及中心庭院的贯通联系，并在建筑靠山体一侧用巨型洞口穿越形体通达内庭。以此形成开放的"吞口空间"。在吞口空间、内庭及建筑主入口一侧结合高差、形体变化，以大量"引桥"的形式联系各使用空间。整个建筑平面整洁、理性、有序，但因引入了地方传统建筑界面类型及空间形态的基因，产生了一种考虑环境特质、有地域内涵的建筑形态。

The layout of the building of the School of Computer Science follows the court-yard style building arrangement, which is a common feature of traditional regional houses. The building side facing the Tainma hill is opened on purpose, creating a direct view connection between the central courtyard and the scenes outside. The concept of 'Connection' within the building moreover has been created in varied forms of approach-bridge in order to integrate various building parts and aid an effective mobility inside. Meanwhile, its facade design has introduced the interface construction and spatial typology of traditional regional buildings, creating a culture connection. As a result, an emergent building responding to the surroundings and regional culture becomes existent.

设计: 魏春雨 罗苊 马迪 宋明星

设计时间: 2008 年

竣工时间: 2010 年

规模: 20562 m²

Architect: Mr. Wei Chunyu Ms. Luo Jin Mr. Ma Di Mr. Song Mingxing

Design start: 2008

Construction end: 2010

Area: 20562 m²

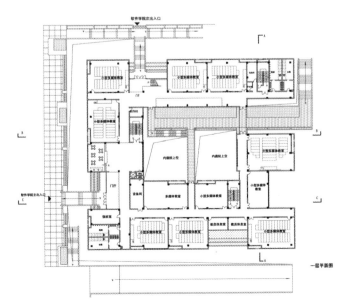 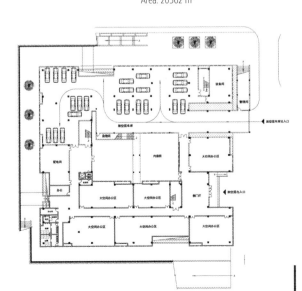

综合教学楼
The Building Complex for Education

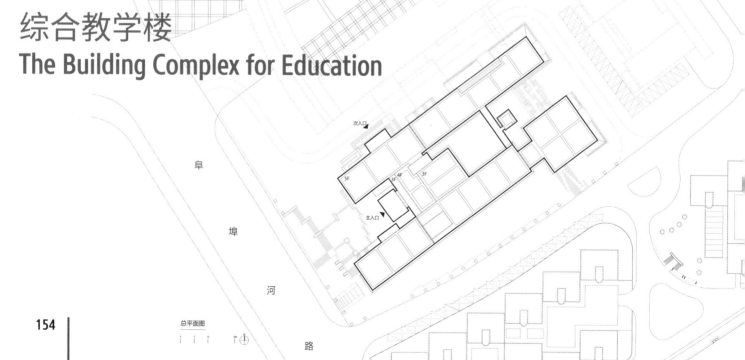

综合教学楼采用线型体结构，以"侧身"形态偏居南端，最大限度地避免对天马山主景区山体的遮挡，并作为山体视觉延展的新的部分。连续线体中嵌入"空中边庭"及类似湘西民居中的"吞凹空间"，中部以"风雨桥"（Covered Bridge）将两部分串联，形成三合院，开放空间对景山体。整个组团仿如聚落，以"群构"形成统一群形态，综合楼部分则是整体群形态的组装部件。

The new development of the university campus could be represented by the newly built educational buildings, the Building Complex for Education. The spatial arrangement gives a good account of the heterogeneous isomorphic design that the area adjacent to the Tianma is designated to educational building cluster with compact design in a sort of layback style. The building is organized in the similar pattern to the regional resident settlement by creating the peripheral courtyards in the vertical dimension with an open view to the hilly skyline of the Tainma hill and utilizing a covered bridge to integrate the different parts of building. Consequently the every part of the building emerges as integral one of the building typology.

设计：魏春雨 李煦
设计时间：2007 年
竣工时间：2008 年
规模：17056 m²

Architect: Mr. Wei Chunyu Mr. Li Xu
Design start: 2007
Construction end: 2008
Area: 17056 m²

游泳馆
Aquatic Center

群构的理念中暗含了地景的思想。各个建筑本体与大地、山体之间是一种彼此嵌入的关系。建筑从大地中生长，大地是建筑的延展，由此消弭了建筑与景观的清晰分界，形成了一种原始粗犷的审美意义。游泳馆位于天马山麓坡地与平地的交界处，形体顺应地形坡起关系，在主入口区域的建筑形体依山就势，层层叠叠似一片片的岩石架设于坡地之上，结合大尺度室外台阶的运用，自然地形成空间场所行为导向，明确建筑与周围环境的地景响应关系，同时也提供了一个容纳各种公共活动行为的平台。建筑场馆主体凌空架设，成为整体构型的重心，好似一个从山体中伸出的方盒，凸显了自身，与周围形成一种对比和衬托，充分展现体育建筑的张力。

设计：魏春雨 黄斌	Architect: Mr. Wei Chunyu Mr. Huang Bin
设计时间：2010 年	Design start: 2010
竣工时间：2016 年	Construction end: 2016
规模：6900 m²	Area: 6900 m²

The cluster configuration of buildings is designed to be dedicated to the landscape architecture, on which treatment of the built ones in this site as a part of environment elements is based. As such, the demarcation between architecture and landscape is annihilated and they become integral nature in the sense of primal art. The indoor swimming pool located at the bottom of the Tianma valley blends into the surrounding slopes via the terrace design, which naturally forms a circulation directing the public to use this space in different purposes. The main body of the building emerges from the surrounding terrace in a big volume to generate the powerful image by this contrast arrangement.

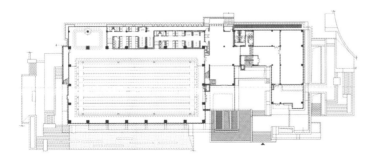 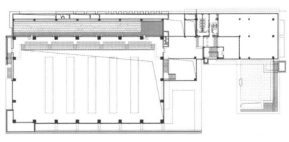

研究生院楼
Graduate School

　　近些年来我们研究了湖南传统民居吊脚楼、风雨桥、晒楼、吞口屋、天井等形式，剥离外在语汇显现特征，提炼和还原出各种基本空间类型，成为指导设计的原型语言，并通过一定的转化衍生在实践（包括在湖南大学原有校区的设计）中大量运用。

　　研究生院楼采用了院落、天井、台地、狭缝等地域空间类型对两万平方米的庞大体量在不同向度进行切割和镂空，消解了建筑对城市空间的压迫，并被塑形成为了一个实体组合感强烈的建筑"雕塑"和内部充满着空隙的呼吸体，这也与湖南大学的建筑原点——岳麓书院的各种空间形态衍生呼应，由此有效地应对了该地点的历史文化、气候环境以及场地特征等各种复杂的制约条件。

After years of intensive study on the regional traditional architecture, a set of building typologies has been sorted as a design language and is deliberately applied on the buildings during the reconstruction of the modern campus, such as stilted building, sheltered gallery bridge, sun-blasted floor design, Tunkou house, light-well, etc. These symbols of language have their own logic connecting to space typologies therefore to function as guidance in both the new and old development of the university campus.

The building complex of the graduate school comprises numerous courtyards, light-wells, terraces and caves space, which have proved to be successful to generate local climate and culture adaptive building space. Meanwhile these elements created in the buildings, for instance "peripheral courtyard", "bamboo light well", " garden in the air", "stack effect well", etc., not only make a holistic urban alike place in such a sculptural beauty style for learning but also echo the design philosophy of the Yuelu Academy-diversity and ecology. To conclude, the design consideration for building complex of the graduate school helps to form heterogeneous spaces for the public and generate dynamic interfaces in the new campus.

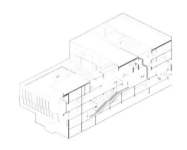 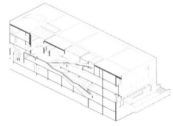

设计: 魏春雨 李煦 宋明星　　Architect: Mr. Wei Chunyu Mr. Li Xu Mr. Song Mingxing
设计时间: 2011 年　　Design start: 2011
竣工时间: 2016 年　　Construction end: 2016
规模: 29012 m²　　Area: 29012 m²

理工教学楼
Science and Engineering Building

设计在楼与楼之间重构了契里柯绘画里深景透视的街景，找寻中间"空"的况味，通过简单的排列逻辑进行组合，产生深邃的、具有某种历史场景感的东西。A楼与B楼面面相对的部分都采用了单元体交错并置的方式，仿佛端坐在两侧的石像彼此之间的缄默对望，中间夹着一条绵长甬道，指向着这条轴线的远端的收口——C楼端部如同弯曲手臂一样的大悬挑体量。

模型推导：改变以功能板块组合的惯常模式，而以类型单元为基础，研究其空间内在的结构关联，形成包含了城市空间、教学组团、单元空间、交通步道体系、视域廊道的群构整合，从而使功能关系组合转为结构关系建构，并以连续单元体强化线性空间特质，强化出契里柯式的"深景透视"的效果，以达到加大景深层次的效果。

The configuration of the three buildings namely A,B,C respectively in the Engineering education premise aims to simulate the streetscape of Giorgio de Chirico's paintings featuring deep view perspectives. This design helps to create an atavistic feeling on the site where people are walking about with the buildings following the simplistic and rationalistic design rules. Building A and Building B sitting along the street of the two sides are facing each other in silence similar to statues standing alone, while the alley wandering through the space in-between of the two buildings stops at the end parts of Building C whose big volume of mass cantilevering in the air like bending arms.

The intention to change the conventional functionalist design is realized by the design approaches of modular and typology. In the same time, the space interrelation of buildings has been created to intensify the deep-view perspective effect of Giorgio de Chirico's painting techniques, and based on that, this building cluster for education is transformed into multi-function urban space consisted of public space, teaching cluster, multi-purpose units, internal footpaths, sight corridors and so on.

设计：魏春雨 宋明星 李煦
设计时间：2011 年
竣工时间：2016 年
规模：A 栋：10378.5 m²
　　　B 栋：6739m²
　　　C 栋：7323m²

Architect: Mr. Wei Chunyu Mr. Song Mingxing Mr. Li Xu
Design start: 2011
Construction end: 2016
Area: A Bulding: 10378.5 m²
　　　B Bulding: 6739 m²
　　　C Bulding: 7323 m²

异质同构的校园
Creating a Heterogeneous Isomorphic Campus

"异质同构"的概念

"异质同构",从广义角度而言,系指不同内涵的事物具有相同的内在结构,是格式塔心理学的理论核心之一。格式塔心理学认为,在人与自然的网络之中,物质的物理活动、人体的生理活动和大脑的心理活动之间,虽然存在着本质的差别,但物、身、心三者在"力"的样式上具有一致的倾向,具有同形的关系,即为"异质同形"。一旦这几种不同领域的"力"的作用模式达到结构上的一致时,就有可能激起审美经验,这就是"异质同构"。正是在这种"异质同构"的作用下,人们才在外部事物和建筑形式中直接感受到"活力""生命""运动""平衡"等性质。

Brief introduction of Heterogeneous Isomorphism

Heterogeneous isomorphism in a broad sense means mixtures with isomorphic composition, and it is one of fundamental theories in Gestalt psychology. Gestalt Psychology have advanced "the heterogeneous isomorphism theory" uniquely after merging many new knowledge, for example, modern physics, phenomenon learning, structure doctrine and so on. Even though the objects, subjects and their surroundings are different in general, Gestalt aesthetics believes that they share a propensity in form of 'force'. Therefore thing, body, and mind relate each other in a similar pattern of force and become isomorphic, namely "Heterogeneous Isomorphism". Because of the aesthetic force of heterogeneous isomorphism, the feeling of "lively", "energetic", "movement", "balance" etc. is able to be aroused during the appraisal of abiotic buildings and some other things.

校园建筑的异质与同构
Heterogeneous Isomorphism of Campus Architecture

从岳麓书院到湖南大学，校园在变化的过程中仍然保持了文脉与空间形态的整体传承，这不是简单的同质复制过程。校园不断地受到新的外部因素的影响与导入，呈现跳跃式、片段式的发展模式，所以它不是一个简单、封闭的过程，而是与内外因素共生的结果。校园建筑的形态与功能都存在巨大差异，但对一些原则的共同尊重，使它们呈现出"同构"的表象。

岳麓书院能够延续千年是因为它是一个"活"的教育场所。如果仅仅作为一个观光游览之地，它就如同一潭死水无法与现代大学的功能融合。而要与之同构，则应关注环境文脉肌理延续、空间尺度协调、严格的空间序列、人文精神传承等方面的融合。

形成湖南大学校园建筑多样化的原因是多方面的，涉及学科设置、功能变异、环境变迁、建造技术更新、新材料应用等因素。而校园空间与之相应，城市与教学场所混合，各个年代、各种风格的建筑并置，它们之间在空间尺度、空间特质、造型语汇、功能构成、材料构成等方面亦存在许多差异。

From Yuelu Academy to Hunan University, the campus has transformed the traditional academy into a modern university of complete different functions and spatial forms. It is an architectural example of mutualistic interaction between the external and internal impacts, between the old and new influences. Whatever the process of transformation, university campus follows a same principle of development, and it appears isomorphic.

After thousand-year development, Yuelu Academy is still a lively venue for education discussions. Meanwhile the campus of the Hunan University is a mix of architecture of different styles over the years. In order to create a campus of heterogeneous isomorphism, the attention should be given to maintaining culture context, coordinating spatial scale, conforming to strict spatial sequence and holding the essence of humanities.

空间尺度的处理

岳麓书院相对低矮、亲和，与山的关系是一种嵌入和匍匐的关系；发展到近代特别是现代，建筑及环境空间尺度趋向于扩大，以解决扩招后及更多通用大空间之需，教学的空间与环境尺度发生了异化。需要解决的问题是：空间尺度变异后，如何延续岳麓书院传统空间的空间肌理。

Spatial scale

The Yuelu Academy is low-story building complex and built in a relative close place at the mountain foot. However, modern education requires airy and open space in order to facilitate varied education and research purposes. It is an issue that the difference in the scale of buildings may break the consistence of the campus texture and create difficulty in keeping the traditional spatial order of Yuelu Academy.

静与动的关系

书院建筑讲究布局匀称，秩序，营造的是相对恬静的书卷氛围。而现代湖南大学与城市空间对接，许多空间节点是喧嚣、流动的。复合的空间与场所的氛围更加趋于动态与不稳定。城市活动以及大量交通的导入，以及作为岳麓山景区的一部分所呈现出的公共性，与高等学府所需的静谧氛围之间的矛盾亟待解决。

The static and dynamic relationship

The layout of Yuelu Academy is symmetric and follows strictly a traditional construction order, hence creating a rather static place with a tranquil and intellectual air. Contrary to it, modern Hunan University is open to the urban space, which is busy and appears a high mobility. The urban life floods into the university campus, and the public space of the university correspondingly turns out mixed use and is unable to define due to its dynamic and unstable state. Therefore it is arguable that the balance relationship between the busy urban life and peaceful learning environment should be the focus of attention of the campus redevelopment.

专属空间与通用空间的分与合

大学常常分为文科与理科等不同学部，是对校园进行理性的归理，形成相对明晰的文理学部分区，还是继续相对散点透视的聚落状分布？这是一个值得研究的问题。同时也要考虑其他附属功能，如教师宿舍、学生宿舍从主校区剥离时会丢失书院教学一体化的氛围，其利弊的权衡不易简单评判。

Exclusive space and general space

Hunan University was a multidisciplinary polytechnic university. After century long development, its disciplinary structure experienced renewal and now covers the academic areas from engineering to social science. It would give rise to a problem how to demarcate area in the campus according to different function needs in space. Whether campus should be defined by two separate parts only used exclusively by a particular disciplinary or maintain one space for general use?

建筑材料的变化

青瓦灰墙—清水红墙—红色拉毛面砖及水刷石。书院的传统建筑材料为青瓦灰墙，而在近代建筑中大量采用清水红砖墙以及水刷石。这些材料的共同特点是朴实庄重，较少雕饰彩绘，色调素雅。材料随时代在变化，但在这些建筑的"血液"中仍然留存着校园老建筑的文化基因，同时具有鲜明的时代烙印。

以上问题的产生，其实为采用"异质同构"理念延续校园文脉提供了契机。

Building materials

Yuelu Academy is a typical traditional academy architecture featuring grey roofing tile and black brick wall. The buildings in the modern times utilize Shimizu bricks and granitic plaster. Although the building materials change over the years from traditional grey tiles with black masonry, Schimizu brick wall to present red wash wall tiles and granitic plaster, the traditional building techniques and construction have been passed down by these campus buildings. It becomes a primer for the campus architecture and helps to create Genius loci in the university campus.

The issues listed as above are the focuses of discussion in creating a campus of "Heterogeneous isomorphism".

"异质同构"的校园建筑设计手法
Architectural Design Approaches of Heterogeneous Isomorphic Campus

校园内的新建筑应当有其自身的美学完整性，应充分表现时代精神。新建建筑应当提升校园的品质并且发展出新的特点，最后达到新老建筑在愉快和谐的氛围里并肩屹立。"异质同构"成为在这样一个校园里进行设计的出发点。我们在设计中持续关注建筑类型与形态的地域性表达以及建筑、校园与城市中的复合界面营造，用现代材料、现代技术来塑造地域化的复合界面。而这种界面不是简单地对校园传统的模仿，也不是简单的符号拼贴，甚至完全没有老建筑的任何形式特征，而是通过类型转换与建筑复合界面来完成"异质同构"，最终达成新老建筑异曲同工的效果。

The development strategies of campus architecture should be made to favor the formation of its new identity with regard to aesthetic integration and then about to reflect the spirit of enterprise of the time. Heterogeneous isomorphism actually embodies the essence of the architectural thinking of several generations of architects who participated in the campus development over the years. Campus architecture of this type focuses on the architectural expression on regionalism and utilizes specific design approaches (for instance, new building materials or technologies) to construct complex-interfaces among the campus buildings, the university campus and the city. As a result, a new campus architecture is able to be created with an identity and merges heterogeneous isomorphic

类型转换与复合界面

转换是结构的基本属性和构成方法之一。转换的最常见方式是在同一类型内的形式变换。由于这种变换是在深层结构基本相似或不变的情况下，表现结构所进行的不同组合，因此又称为"基本转换"或"类型转换"。转换或重构利用重复、变形、分离、叠置等转换手段，实现内与外、虚与实、逻辑与形象、偶然与必然、片段与整体、理性与浪漫的有机重构。在校园新地域的实践中，运用"类型转换"与"复合界面"两种不同的形态处理手法，以期达到新老建筑和谐共生的效果。而从岳麓书院到湖南大学一批优秀的近现代建筑中，建筑师提取相当多的形式类型语汇进行转换，如内置天井、清水砖墙、水平向度的处理、坡地台地建筑处理等，有机融入到新建筑中去。

Typo-transformation and complex-interface

Transformation is a term generally referring to a constructive method for form alternation of structures categorised in the same group. This alternation occurs under condition that the underlying structures stay same and the changing process of this type is specified by a term, namely 'basic transformation' or 'typo-transformation'. A number of measures can be applied for 'typo-transformation', such as duplication, deformation, disaggregation, juxtaposition etc. In the practices of regional architecture in the campus, the chances to reconstruct the relationship between the inside and outside, virtual and real, logics and form, contingency and necessity, fragment and totality, rationalism and romanticism could become possible by the implement of "typo-transformation" and "complex interface". Therefore the harmony state between the old and new buildings could be achieved using typical building elements when transforming form, for example light well, Shimizu brick wall, implication of horizontal dimension, terraces, etc.

建筑复合界面具有层次性、连续性、多义性。它是建筑与城市之间过渡性的层次，具有空间与界面的共同特性，并包含了围护实体、空间、人文活动等三种要素。校园复合界面是校园设计过程中建筑组群及单体有机复合形成的整体，它涉及：建筑单体及群体与校园空间作用产生的界面，校园外部空间环境界面及人的行为、感知心理模式与校园建筑界面的互动影响。由于是在老校园中穿插式地进行新建及扩建，常常遇到新旧并置的情况，如何解决两者在空间尺度、造型语汇以及功能上的差异是棘手的问题。我们结合湖南冬冷夏热的气候特征，以及延续传统校园的空间文脉，尝试了"边庭""竹井""空中边庭""风拔"等空间塑造手法，营造了复合化的建筑形态。

Complex interface in architecture is of multi-layer, succession and ambiguity. It is the transition domain between the buildings and their surrounding urban space with a certain level of similarity in the spatial typology and construction pattern. In addition, it basically encompasses three layers physical building envelope, spatial form, and place-making. For complex interfaces created during the course of campus designing, particular attention must be given to effective interfaces among a building, building complex, and their surroundings in the campus, to the external space form of the campus, and to the space users' behaviours and perception modes responding to these complex interfaces in the campus. Innovative measures have been experimented in the campus development, and some techniques, for instance "peripheral courtyard", "bamboo light well", " garden in the air", "stack effect well", etc., have proved to be successful to generate local climate and culture adaptive building space.

异质同构的校园新建筑

我们尝试以类型形态的地域表达及复合界面的营造来指导设计实践，并形成我们的共识：地域性的形式类型可以进行有机转换；可以用片断或局部（如地域材料）来诠释地域性，无需背负全部地域符号和文脉；复合化的界面利于调和校园新旧建筑的差异，是一种与环境共生的方法；复合空间与界面为融合地域文化、调节生态小气候提供了可能。

湖南大学生物技术综合楼位于岳麓山南大门及东方红广场南侧，毗邻蔡泽奉先生与柳士英设计的原湖南大学科技馆。设计上没有简单地采用"同质复制"的手法照搬相邻科技馆的式样，而是通过"异质同构"的方法，采用现代建筑风格，并将其作为"配角"，以一虚一实、一藏一掩的造型与书院精神求得呼应。

New campus architecture of Heterogeneous isomorphism

Until now, attempts on construction of a new campus have been given an account by regionalism architectural interpretation and complex interface generation. Regionalist methods help to strengthen the historical links between architecture of different periods and bring out a culture identity to the place where the architecture has taken root. The practices like local material implementation, traditional construction techniques pick-up are useful to create regional campus architecture. While the complex interface concept is beneficial to the architecture discourse between the old and new campus as well as microclimate regulation. Building of Bio-tech in the Hunan University is situated in the central area of the campus, next to the Building of Science and Technology, designed by Mr. Cai Zefeng and Mr. Liu Shiying. The design approach of the Bio-tech Building was not to simply replicate the style of its neighbours but to carry out heterogeneous isomorphic design, based on which the building was designed to act a supporting role in such a 'stage-set' of campus .

在复临舍教学楼的设计中，通过采用入口灰空间、边庭消解柔化了立面，强化了室内外环境的交融；在材料上采用灰红色的亚光面砖。该建筑打破了传统单廊及内廊教学楼模式，以内厅、边庭、交往空间形成集交通、疏散、交往、展示等多功能活动空间。

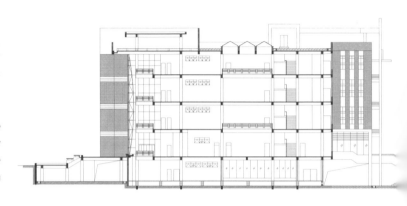

Building complex of Fulin emphasizes on the communication between the inside and outside by creating "grey space" at the entrance, utilizing periphery courtyards and adopting open corridor arrangement along with a soft facade design. The mobility within the building has therefore been enhanced, and then a flexible and open place emerges for varied functional requirements.

在工商管理学院的设计中，我们的设计理念基于对基地中央的一棵大樟树的保护。将树作为中心，以大片实墙将周边喧嚣烦扰阻隔，西立面挖出三个竹庭，软化实墙带来的绝对封闭，同时北立面使用双层界面，在建筑与道路间形成一个过渡层次。通过对建筑形体与环境界面的复合化处理，形成相对宁静的学院氛围。

The design concept of the Building of the Business School was inspired by the attempt to conserve a camphor tree, standing in the centre of the site. The building is arranged around the central tree that it is open to the centre while the solid wall at the west wing is designed to keep the quiet indoor environment out of some interruption from the street on the west border of the site . Concepts of complex interface have been applied to the elevation design of its four sides.

法学院、建筑学院建筑群考虑与周边校园环境的协调，寻找自身与湖大内部文脉的联系，保持二者间形式上的延续与完整。建筑在材料的选择与形态的设计上采取同场地相一致的原则，从建筑本体到场地自身限定的区域都采用了水刷石饰面，并设置大量舒缓的台阶，形成完整的"场域"。建筑融入场地，就像从土里生长出来的。

The building complex for the School of Law and School of Architecture seeks for the architectural succession of tradition and integration in the campus. Granitic plaster has been applied in a large quantity on the wall of the building in order to create a material identity of the place. Terraces gently run down along the building's north and west wings. A building responsive to the site where it grows out of takes into shape .

而在天马新校区的设计中,我们抽取书院建筑理性有序的格局,并采取整合与群构的形态以及与地形融合的策略。在毗邻天马山的区域,建筑形体尽量消解、复合化,而远离天马山则采取群构的空间肌理与形态,形成地景建筑。场地最南端的综合教学楼及最北端的软件学院的设计遵循整个新校区的规划理念,形态上没有做过多复杂的变化,而是将自身作为群构网络体系的一个组成单元,为整个组群的有机生长做好准备。复合化的建筑形态形成了性质各异的围合空间。入口广场是外向的,开敞的,动态的,通过不同标高的变化来丰富外部环境;建筑内部庭院则是内向的、安静的、相对私密的,通过大的台阶、平台、连桥来实现内部、外部环境的立体化。内院两侧的形体如同画框,将对面的山景框入画中,形成一个天与地相交的界面。

The new development of the university campus could be represented by the newly built educational buildings in the Tianma valley along the new planning axial line of the campus. The spatial arrangement gives a good account of the heterogeneous isomorphic design that the area adjacent to the Tianma is designated to educational building cluster with compact design in a sort of layback style, i.e. the Building Complex and Building of the School of Computer Science, while those areas on the other hand away from the Tianma are dedicated to the landscape architecture. Complex building typology helps to form heterogeneous spaces in the site and generate dynamic interfaces in the new campus .

场所的语义：
从功能关系到结构关系
—— 湖南大学天马新校区规划与建筑设计

近年来，大学校园的建设方兴未艾并取得有目共睹的成果，但快速的发展过程也显现出诸多弊端：一味追求功能的高效和分区的明晰使得校园的有机性、灵活性和可识别性丧失殆尽，校园应具有的礼仪感和学府精神被直白、机械的追求效能的组织关系所取代；校园的文化传承成为口号，很难找寻到可以依附的物质空间载体；校区与城市和环境肌理的对接往往是生硬的隔离；各种看似无比正确的设计准则和规范的导引下获得的结果却往往是空间语言的苍白和乏力。

湖南大学天马新校区的规划建设让地方工作室获得一个破解上述问题的难得机缘。新校区位于岳麓山大学科技园，总用地 18.96hm²。场地北临湖南大学老校区，东倚天马山，往西隔麓山南路与岳麓山相望，南面为阜埠河路和天马学生公寓。校区从规划设计到全部建成历经 15 年，各种内外部的因素导致了校区漫长的、跳跃式的建设周期，这也使得设计主体有了主动和被动的、不断审视自身的可能，在此过程中，相应的设计思路也逐渐发生了多重蜕变，从对功能和环境的制约条件被动适应，到地域类型的自觉运用，再到多年沉淀心理图式的自然浮现，最终体现为他律、自律、自在的不断转化同时又彼此共存，这些思考的背后是以结构主义语言学理论和受结构主义影响的理性主义类型学理论为支撑。于此，我们突破了功能主义的禁锢，转而聚焦于群体表象下的深层次的结构关系。

1 软件学院大楼　　5 运动场地
2 研究生院楼　　　6 理工楼A栋
3 游泳馆　　　　　7 理工楼B栋
4 导航塔　　　　　8 理工楼C栋
　　　　　　　　　9 综合教学楼

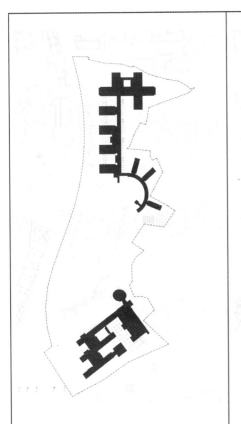 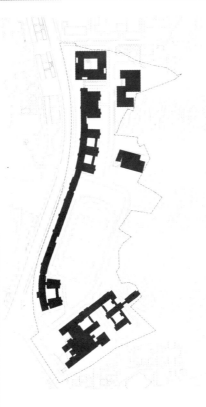 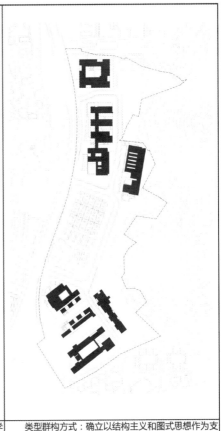

| 鱼骨行列方式：试图结合功能与场地关系建立一个整体新型校园的概念，以公共廊道串联各个教学单元，总体架构关注功能关系，采用强调各教学组团整体联动鱼骨式的排列方式，从而对功能和环境的制约条件被动适应。 | 围合类型方式：设计过程提出类型的概念，把教学属性更加具体化，分别梳理出综合版块、艺术版块、理工版块、研究生院版块，简单强调了屏蔽与围合，形成隔绝城市喧闹街区的大围合的形态，属于地域类型自觉运用的尝试。 | 类型群构方式：确立以结构主义和图式思想作为支撑，化解常规的功能束缚，通过群构与地景策略的使用、地域类型的衍生、仪式化空间的营造来生成外在与内在的场所语义。设计操作按照学科属性进行组织，建构出相应的秩序和逻辑，保证了功能使用的高效和便捷，功能关系虽会被考量，但个体与群组之间的关系才是关注的核心。 |

1 湖南大学校区的整体发展脉络

  天马新校区的规划设计（包含游泳馆、理工楼、研究生院楼等），设计必须置于湖南大学整体的时空关系下进行考察。湖南大学校园的发展流变有几个交织的维度：一是岳麓书院的历史文脉，这个维度是书院历经千年发展而来的，书院的空间形制是湖南大学空间传承的一个重要基因和原点；另外一个维度是柳士英、刘敦桢、蔡泽奉等先贤主持规划建设时期遍布校园多处的近现代保护建筑，包括以红砖作为建造和装饰材料的老图书馆、科技楼、教学北楼、大礼堂、胜利斋、二舍、七舍等，这些建筑已经构成了湖南大学近现代发展的一个基本的风貌格局，其后到1980年代还建设了电气楼、化工楼等以水刷石为外墙材料的朴实校园建筑；第三个维度是湖南大学地处岳麓山国家风景名胜区，校园的规划建设一直是和山地景观交织在一起，这既是特色条件，也在一定程度上存在着某种刚性的制约，基于客观情况，天马新校区的建设有别于当今很多学校一蹴而

就、一次成型的轴线序列和空间形态的建构方式，设计需结合地形坡势，处理不规则的、犬牙交错的场地边缘，甚至是一些不规则、线性的空间。上述情形造成了湖南大学校园的这种历史递进性和与环境共生的发展状况，因此校园的建设也不会按照一个自上而下的设计贯彻到底，而是一个不能完全被"设计"出来的校园，它需要不断地和周边环境磨合、交接、剥离，形成一种整合共生性的存在状态，这也是湖南大学一种独特的存在方式，同时也无心插柳般实现了现代大学精神所在——开放性和多义性。所以整个发展过程事实上也契合了岳麓书院空间完整的演变机制——由岳麓山脚下青风峡引出，经自卑亭、穿越凤凰山与天马山之间的牌楼路、面朝湘江和长沙城区而形成一条千年古文脉。在这个过程中书院空间一直与城市具有相对独立性，彰显了学府的精神，同时跟城市与自然环境是一种契合的状态，因此有"纳于大麓、藏之名山"的典故。这样的特征也必然成为新校区建设的指引和启示。

特殊的时空关系决定了湖南大学的校园空间发展面对的问题不是去营造一个单纯的教学场所，校园的建设发展需要一个新的思考角度。多年来身处其间、作为设计者的我们对这种模式深有感悟与体会：这已经不是一个单纯的协调与融入，而是如何将大学的形制和空间精神同环境特征结合在一起。因此，要怎样实现异质同构和新旧并置成为校区规划设计核心的考量。曾有一个过渡时期，地方工作室在湖南大学核心区范围内设计了生物学院大楼、工商管理学院大楼、复临舍教学大楼。如果说岳麓书院是"灰调时期"，那个时期就是属于"红调时期"。此时，由于建筑的体量、所处的区位，设计更多的是顺应核心校园区的总体的场所氛围。新校区的选址不在核心区范围内，书院和近现代建筑群不会构成直接的影响。设计面临的问题转化为如何处理3个关系：如何协调好与天马山景区的关系；如何处理好与毗邻基地的麓山南路商业街的关系；诸学科功能如何调配。同时，新校区规划设计在湖南大学发展的整体空间结构上肩负这样一种责任：将校园空间向南拓展形成一个新的轴线——从岳麓书院、到大礼堂、到图书馆、到朱张渡口的与文脉轴线相垂直的教学轴线。在这些之上，地方工作室想追寻一种深层的结构，实现自治性与他治性的结合。

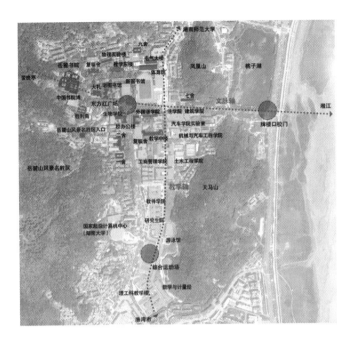

2 新校区规划建筑设计的理论支点：结构概念与图式思想

新校区的规划设计所围绕着的核心是要建立一种呼唤精神回归的认知和设计价值体系，也就是地方工作室多年关注和研究的、在阿尔多·罗西的理性主义类型学的基础上所塑造的一种新的场所语义，其理论支点来源于结构概念和图式思想。

皮亚杰（Jean Piaget）对结构的概念做出了定义，他认为：人们可在一些实体的排列组合中观察到结构，这种排列组合体现了整体性、转换性和自我调节性3种基本特征。结构主义者的思维方式的基本准则就是：在任何既定的情境里，一种因素的本质就其本身而言是没有意义的，它的意义事实上由它和既定情境中的其他因素间的关系所决定。地方工作室通过一种形而上的逻辑操作方式，摒弃了简单的功能集合方式，把新校园的诸多建筑纳入一个统一的形态衍生秩序的背景之下，关注的对象不是低层次的呼应关系，而是使用者的深度体验。设计把各个建筑单体纳入到一个统一的格构组织之下，考察它们构成的组群之间，组群的局部之间以及组群局部和整体之间形成的

空间对话与关联。校园被视作一个结构体，每一栋楼自身的功能、空间、风格等外显特征在整体中是被"抹去"的，它们之间的互动同构关系才是探究的重点。

在整体格局下，每一个体的空间形式却具有多义性，这些繁杂的表象被一种深层的逻辑结构所支配。这一结构与地方工作室所追求的建筑自治性相关。工作室近些年来持续关注地域的现代性与现代性中的表达问题，发掘深藏在事物表象下的内在逻辑：深层结构—类型原型—心理图式，据此提出了"地域类型图式"来回应当下的地域创作的方向性问题，其中涉及图式内源性（形式的内在演变规律和基础）与外源性（形式变化的外在制约条件）问题。内源性具体包含了"现代艺术""周边式构图"及"新的结构体系"，这些都是属于现代建筑的学科内部问题；外源性则包括了地形、文脉和场所精神等。内源性中的现代艺术是设计创作内核最重要的来源，具体来说就是契里柯（Giorgio de Chirico）与卡洛·卡拉（Carlo Carra）于1917年创立的"形而上绘画"，他们的画面把真实与非真实犹如缠绵的梦境融合在一起。形而上艺术表面上十分宁静，但给人的感觉却像是在宁静中会有什么事要发生。建筑师阿尔多·罗西的绘图和作品显露出相似的气质，这体现在两者都强调通过运用哲学方法让人们产生某种抽象的意识，而不仅仅是感官上看到的具体的物象，并且两者作品呈现出相同的图式语言特征，即差异并置、扭曲空间、几何简化、阴影渲染，以此来实现契里柯式的如梦境般的视角。这些图式语言在校园建筑环境营造中有所体现，也由此形成了校园自身某种带有神秘主义的特殊气场。

3 场所语义的建构

新校区既是教与学的功能化场所，也是设计者思想的一种投射。在这里，思想的物化具体是通过群构与地景的形态、地域的类型以及仪式化氛围来实现的，并以此获得多重的语义。

3.1 群构与地景

新校区毗邻老校区，是它的延伸、扩展和补充，设计以

群构的方式组织了各个功能体。群构是一种构建连续、完整的建筑和城市形态的方法，其核心是组群公共空间形态的塑形，追求的是建立在结构主义思想和格式塔完形理论基础上的一种空间整合性。规划设计中，由北至南的道路成为了主脊，连接了软件学院大楼（包含艺术设计学院）、研究生院楼、游泳馆、运动场、理工楼（包含A、B、C三栋）、综合教学楼等建筑与场地，也同老校区的法学院、土木学院、工程实验大楼相对接而形成了湖南大学校园完整的教学轴。各建筑使用者的教学、科研、办公、实验、运动等行为的差异导致了功能内容纷繁复杂，尺度规模大小不一。设计操作按照学科属性进行组织，建构出相应的秩序和逻辑，保证了功能使用的高效和便捷，如将物理楼、微电楼和数学楼整合为一个群组。在大的分区原则下，设计也注重不同学科的交流融合，如将艺术设计学院和软件学院置入到一栋楼内，以期产生跨学科思想碰撞的灵感火花。功能关系虽会被考量，但个体与群组之间的关系才是关注的核心。在关系处理上，设计不过多着墨于单体的形构特异性和自身属性的强力彰显，谨防校区成为个性化建筑的杂烩拼盘。单元衍生构成策略的采用在各单体之间以及单体各部位之间形成对位、呼应、相似、穿插关系，保证了它们在尺度上的协调，形态的划一，加之一种外墙水刷石材料的纯粹使用，让各个建筑集聚为整体。形体连接、抬升、挤压、切割、退缩操作后留出了基座、平台、院落、廊桥、巷道、天井等空间。这些空间散布各处，成为促进人际交流和激发更多创意的场所，它们突破本有功能定义，形成了新的意义延伸。另一方面，这些空间同时产生了阴影区、风道，也是对本土地域气候的一种积极回应。群构的建筑还涉及与城市道路边界接触问题。操作时，设计有意拉开一定的距离，化解了城市道路的放坡斜度与基地的斜度不一致的矛盾，相应地减少了对东侧山体的遮挡，留出人观看建筑的空间，结合草坡、下沉、连桥在校区和城市之间构成了一个柔性过渡。新校区呼应和匹配了湖南大学原校区无围墙的做法，展现了"海纳百川、有容乃大"的自信和气度，同时依靠边界处理和整体建筑构型气质的把控来定义出自我的独特存在。面对这些年学校已经被城市道路割裂的现实，设计呼唤一种书院和学府精神的回归，所以采取了和环境并置的方式，

建立起一个自我的体系，以此对周边世俗商业氛围构成一种无形的抗拒和疏离感，并强调大学校园在纷乱的城市以及复杂的制约条件下自我存在的价值。

群构的理念中暗含了地景的思想。各个建筑本体与大地、山体之间是一种彼此嵌入的关系。建筑从大地中生长；大地是建筑的延展，由此消弭了建筑与景观的清晰分界，形成了一种原始粗犷的审美意义。以游泳馆为例，项目位于天马山麓坡地与平地的交界处，形体顺应地形坡起关系，在主入口区域的建筑形体依山就势，层层叠叠似一片片的岩石架设于坡地之上，结合大尺度室外台阶的运用，自然形成空间场所行为导向，明确建筑与周围环境地景的响应关系，同时也提供了一个容纳各种公共活动行为的平台。游泳馆主体凌空架设，成为整体构型的重心，好似一个从山体中伸出的方盒，凸显了自身，与周围形成一种对比和衬托，充分展现体育建筑的力度感。

3.2 类型的衍生

新理性主义类型学家阿尔多·罗西在当时主流的建筑浪潮下，仍然执着于对少数永恒建筑形式的追求，在荣格原型理论、

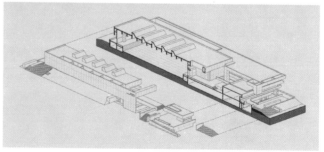

二元性结构主义思想以及现象学影响下，提出了建筑形式具有自律性的观点，建立了自己的抽象类型学思想。在罗西的观点里，类型构筑于美学渴求之上。他认为类型具有某种恒久性，不会受到社会、形态、功能变迁的影响。后来的类型学家将它总结为功能类型与形式类型可以分离，类型是固定的、形式化的框架和容器，至于里面放置什么样的功能空间是次要的。这样的观点是对现代主义建筑过于强调"形式追随功能"论调的思辨性批判，强调建筑不是一个对着千变万化的环境产生反射的镜面，它不会对经济、社会、文化和技术上的事件亦步亦趋，它封闭自身，有着固定演化的秩序，由此体现了建筑形式的自治性以及建筑学科的独立性。

近些年来我们研究了湖南传统民居吊脚楼、风雨桥、晒楼、吞口屋、天井等形式，剥离外在语汇显现特征，提炼和还原出各种基本空间类型，成为指导设计的原型语言并通过一定的转化衍生在实践（包括在湖南大学原有校区的设计）中大量运用。这样的做法在后续使用体验中也获得了良好的反馈，新校区建筑设计仍然沿袭了这些空间类型，并结合基地的实际情形如地形、景观、周边建筑等条件进行了拓扑变形，可以说新校区设计是我们各种地域类型语言运用的集大成者，表达了对建筑的本源和环境属性的全面思考。如研究生院楼就采用了院落、天井、台地、狭缝等地域空间类型对 2 万 m² 庞大体量在不同向度进行切割和镂空，消解了建筑对城市空间的压迫并被塑形成为了一个实体组合感强烈的建筑"雕塑"和内部充满着空隙的呼吸体，这也与湖南大学的建筑原点——岳麓书院的各种空间形态遥相呼应，由此有效地应对了该地点的历史文化、气候环境以及场地特征等各种复杂的制约条件。

类型的使用是对建筑本体和局部的操作，它们不应成为碎片化的孤立的点，需要以群构的方式将它们链接关联，形成联动，共同组织进一个有机的系统中。在该系统中，实体与空间彼此咬合镶嵌，形成了一种致密的三维化肌理（不同于一般所指的

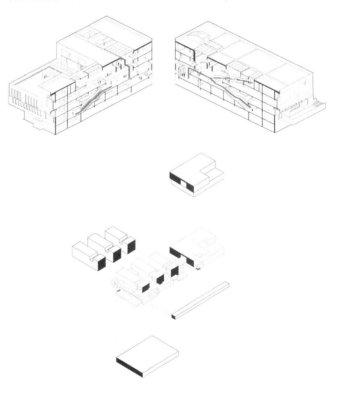

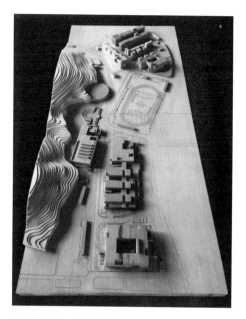 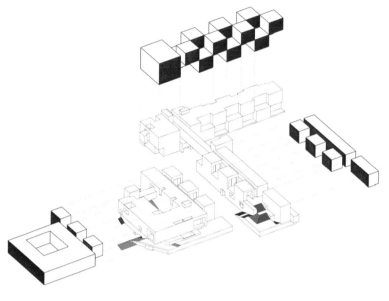

平面肌理），其中叠入不同使用人群的各异行为，一种多层级、各向度和动态性的空间网络便被编织出来，由此校区整体彰显出如同传统民居聚落一般的绵延和有机的效果。

3.3 图式的再现

在地域类型多年的研究与实践的基础上，我们进一步关注场所的内在语义，这种语义与人的情感体验相关。基地西侧毗邻喧嚣的麓山南路，这恰好给予设计一个制造闹中取静、相对独立的场所语义的充分理由。新校园应是修心的道场，有着自己的风骨，我们刻意塑造出如契里柯形而上作品的气氛——画面静寂而悠远、房屋封闭而光影浓烈（在罗西的方案图也流露出相同的气质）。氛围的构建通过两种途径来实现。一种是深景透视的运用。它来自于契里柯作品的启示，他偏好一点透视，突出纪念性，但同时有意在同一画面中采取不同的透视灭点，强调场景的纵深感和扭曲感，偏离人的认知常规，从而传递出绵延不尽的距离感和神秘寂寥的形而上格调。在楼与楼之间，设计重构了契里柯绘画里深景透视的街景，找寻中间"空"的况味，通过简单的排列逻辑进行组合，产生深邃的、具有某种历史场景感的东西。其中最典型的是理工楼群的设计，A、B楼（物理与微电子科学学院楼）面面相对的部分都采用了单元体交错并置的方式,仿佛两侧端坐着石像彼此之间的缄默对望,中间夹着一条绵长甬道,指向着这条轴线的远端的收口——在 C 楼（数学学院楼）端部如同弯曲手臂一样的奇怪悬挑。另一种途径就是墙面水刷石材料的运用。采用这种材料有着历史文化传承的考量——老校区充满着使用这种材料的新旧建筑，但更多的是对石子自身特有属性对塑造氛围所起作用的认识。上墙的天然的卵石是从湘江中淘出、并经过人力的多次筛选而来的。如果仔细辨识，可以发现每颗石子形状各有不同，颜色在同一色系下也千差万别，石子之间甚至可以看到嵌入的贝壳、生物残骸和砂砾等杂质。这无数差异化的微表情在光线的沐浴下显现成一种浓烈的暖调，强化了建筑的棱角和体块，弥散在校园的场所中，人们都被笼罩其中，还原了契里柯的画意。在两种途径之外，一些局部的处理也继续强化了校园仪式化的氛围，例如通过狭缝空间中雕琢出的线状光形去表达具有未知感的阴影领域；局部矗立的框架暗示门的意象和不同领域转换的分界；研究生院楼屋顶平台上看似突兀地被置入旋转 45° 的露天舞台；一些非理性挑出的实体成为"布道台"；建筑入口设置为以桥连接表达与世俗空间的疏离；各处地面隆起的平台预示智者要于此讲演；保留原生的两棵大树形成局部空间领域的指向中心，于此，日常的学习行为似乎变得神圣。校区建筑和

环境中没有一个具象的符号，却似乎处处充满着隐喻，场所内弥漫的意义不断促发人的情绪的喷涌，人在其中不自觉地会被涤荡身心，褪尽浮华，成为一个个浸入知识汪洋的"修行者"。

结语

国内大学校园设计在功能布局、空间形态和建筑形式方面已经做出了有益的探索，而地方工作室关注的重心是去探究一种深层次的、形而上的场所语义，这与湖南大学校区独有的时空特征（包括源自岳麓书院的悠长历史、得天独厚的自然环境、异于常态的城市区位关系等）相关。地方工作室早期在核心区的建筑设计是采取同质异构的方式，取得了融入式发展的效果。其最终的新校区规划建筑设计是功能关系上的进一步延展，在结构思想的指引下实现了内外场所语义的交织与发展，也传达出一种原朴的场所精神。

新校区的建筑有效地应对了气候特征、功能效率、人际交往、两侧山体景观的视线要求以及基地与城市道路的关系等外部制约条件，是设计他律的体现；从湖南传统民居中提炼出各种空间类型的反复运用成为笔者的自觉和自律。场所的外在语义就在这种他律和自律的不断转换和反复叠加中逐渐显影。

每天，在某些时间点，人们总是能看到这样的景象：外部穿梭交织的人流被沉静的建筑收纳，消散不见，此时校园建筑整体氛围如同层层的滤网，滤掉了人内心的躁动欲望，还原了一个清澈澄明的世界，而这就是设计所追求的内在的场所语义。

参考文献

[1] 曹俊峰. 论康德的图式学说 [J]. 西方哲学, 1994(6): 50.

[2] 特伦斯·霍克斯. 结构主义和符号学 [M]. 瞿铁鹏, 译. 上海：上海译文出版社, 1987: 6-9.

[3] 魏春雨, 刘海力. 图式语言——从形而上绘画与新理性主义到地域建筑实践 [J]. 时代建筑, 2018(1): 190-197.

[4] MONEO Raphael. On Typology[J]. Oppositions, 1978(13): 22-45.

[5] 裘知. 阿尔多·罗西的思想体系研究 [D]. 哈尔滨：哈尔滨工业大学, 2007: 53, 68.

[6] 魏春雨. 地域建筑复合界面类型研究 [D]. 南京：东南大学, 2011.

[7] 魏春雨. 地域界面类型实践 [J]. 建筑学报, 2010(2): 62-67.

[8] 魏春雨, 李煦, 杨跃华. 设计的逻辑——湖南大学综合教学楼设计 [J]. 建筑学报, 2011(10): 88-89.

[9] 韩亚飞. 契里柯绘画研究 [D]. 保定：河北大学, 2012: 9, 17.

[10] ROSSI A. An Analogical Architecture[M]//NESBITT Kate. Theorizing a New Agenda for Architecture: an Anthology of Architectural Theory, 1965-1995.

湖南大学校区总平面图
Site Plan, Hunan University

附录 Appendix

湖南大学校园重要建筑一览表

建筑信息	二院（现物理系实验楼） 1925年建成	原图书馆（1938年被日军炸毁） 1934年建成	二舍（现研究生院） 1935年建成	科学馆（现湖南大学校办公楼） 1937年建成
建筑外观				
设计	刘敦桢	蔡泽奉	柳士英	蔡泽奉 柳士英
建筑信息	九舍 1946年建成	胜利斋（教师宿舍） 1950年建成	六舍（现一舍） 1950年建成	七舍 1951年建成
建筑外观				
设计	柳士英	柳士英	柳士英	柳士英
建筑信息	工程馆 1953年建成	大礼堂 1953年建成	老图书馆 1954年建成	岳麓书院修复与设计 1982-现在
建筑外观				
设计	柳士英	柳士英 黄善言	柳士英	杨慎初 黄善言 柳肃等

湖南大学校园重要建筑一览表

建筑信息	图书馆 1981 年建成	环境工程馆 1986 年建成	体育馆 1998 年建成	复临舍（综合性教学楼） 2000年建成
建筑外观				
设计	巫纪光　黄善言	李继生	巫纪光	魏春雨　邓毅
建筑信息	生物技术综合楼 2002 年建成	研究生院逸夫楼 2002 年建成	机械与汽车工程学院 2003 年建成	法学院、建筑学院建筑群 2004 年建成
建筑外观				
设计	魏春雨	杨建觉	柳展辉　罗朝阳	魏春雨　宋明星　李煦　齐靖
建筑信息	机械及汽车学院实验室 2004 年建成	工商管理学院楼 2006 年建成	综合教学楼 2008年建成	软件学院楼 2010 年建成
建筑外观				
设计	魏春雨　李煦　齐靖	魏春雨　宋明星　李煦　齐靖	魏春雨　李煦	魏春雨　罗荩　马迪　宋明星

湖南大学校园重要建筑一览表

建筑信息	超级计算机中心 2010年建成	书院博物馆 2014年建成	研究生院楼 2016年建成	游泳馆 2016年建成
建筑外观				
设计	杨建觉	魏春雨 齐靖	魏春雨 李煦 宋明星	魏春雨 黄斌

建筑信息	理工教学楼 2016年建成
建筑外观	
设计	魏春雨 宋明星 李煦

湖南大学的今昔

柳士英

湖南大学坐落在长沙市湘江西岸的岳麓山下，前临湘江，后倚麓山。远在九百多年前，宋朝时候，这里就创建了一所岳麓书院，作为封建时代文人学士讲学之所。宋儒张栻和朱熹都在这里讲过学，当时学生超过一千人。朱熹所写"忠孝廉节"四个大字的石刻，现在还保存在校内。清朝末年，书院改为近代高等学校。1926年湖南大学成立，到现在也有三十四年的历史了。

解放以前，湖南大学虽然办了二十三年，可是由于校舍不多，图书仪器设备简陋，在校学生平时不过五百人，校里也设立了文、理、工、法等几个院系，但有的系有时一班只有一、二个学生。教职员也因生活贫乏，无心教学。

新中国成立后，随着祖国文化教育事业的飞跃发展，湖南大学和其他院校一样发生了巨大的变化，为适应全国，特别是湖南工农业经济和科学文化事业发展的需要，已成为一所新型的文、理、工合一的综合性大学。现在全校共有十一个系三十一个专业。现代科学新成就，如运算技术、原子能科学、同位素、半导体以及各种新产品新技术，都设置了专业。各专门学科都有实验室、模型室、资料室等，为教学、科研、劳动生产全面锻炼提供了有利条件。

随着师生员工人数的增加，学校的基本建设也迅速发展。校区面积由三、四百亩扩展到一千数百亩，校舍建筑超过二十万平方公尺。有藏书三、四十万册的图书馆、有可容三、四千人的大礼堂，南北分列的各个教学大楼，都是几千平方公尺的大建筑物。在校区边缘的平地和山坡地带，遍建师生员工的宿舍，宿舍四周，满种花木，环境清幽。全校房屋，组成了一个壮丽瑰玮的场面。解放后，河西设了专用自来水厂，引水上山，家家户户有自来水和卫生设备。由河东电厂架线过河，日夜供电。福利设施有各种类型的食堂、有医疗卫生机构、有托儿所、幼儿园、有万能服务站，生活日用需品，不须过河采购。

湖南大学是综合大学，主要培养科学研究人才、高、中等学校师资和有实际操作技能的高级工业建设人才。现有学生六千五百多人，比解放前增加十倍多。学生来自湖南、广东、湖北、河南、江西、广西各省，都是祖国的优秀青年。他们不但有安定的学习环境，在物质生活方面，也得到国家的充分关怀和照顾，全部学生享受国家公费医疗的待遇，百分之八十的学生取得人民助学金。台湾军政人员的子弟董孝论（董后来成为柳先生的研究生，毕业后分配到浙江省建筑设计院工作，1993年被评为全国优秀设计院院长——编者注）等都得到助学金，学习、生活得很好。所有学生，在"使受教育者在德育、智育、体育几方面都得到发展，成为有社会主义觉悟有文化的劳动者"的总目标下，个个顽强地学习科学、积极地锻炼身体，不断地提高政治觉悟。他们懂得：今天的学习，就是为了明天的社会主义建设；因此，热爱学校，热爱专业，信心百倍地为祖国的需要和自己的全面发展而努力着。

由于学校执行了教学必须与科研、劳动相结合的方针，在1958年全国生产大跃进的形势下，在校内设置了十几个生产组织，获得了主要研究成果数十项。劳动是结合专业进行的，要设计,也要施工;要独立思考,也要独立工作。这样,就改变了过去纸上谈兵的搞毕业设计的方法。例如土木系,承担了许多房屋的设计,有涟源钢铁厂、长沙新火车站、沅江人民公社、国际饭店、歌舞剧院、长沙铁道学院等。在自己校内,从设计到施工,兴建了一栋四层无眠空斗墙的教师宿舍,一栋十八公尺跨度没有拉杆的砖砌双曲拱新技术大厅,这两个工程都是国内外首创,具有科研的重大价值。同时,还能结合地方材料,首创各种竹结构样式、玻璃丝磷酸盐混凝土、白云石塑料工艺品、水道用缸瓦管等产品。机械系成批地从事机械制造,接受了本省交下来的机械生产任务。至于毕业学生,分布在全国各地担任各种不同的工作,都能用其所学,为祖国社会主义事业贡献智力。

湖南大学的教职员工,解放后都得到优越的工作条件和优厚的生活待遇。如朱阶平教授,现在土木系任教,一方面担任固定的教学任务,同时还做了一些科学研究工作,他在给水排水工程上很有研究,成功地完成了"多层多格沉池",使用效果很好,有着经济上的重大意义。铁道工程的老前辈桂铭敬教授,担任铁道建筑系主任(该系和有关的另外两个系,今年从湖大分出去,成立长沙铁道学院),还能带着学生四处实习,干劲十足,他的子女都已长大,担任工作。全家生活工作得很愉快。其他如化工系钟龄教授、物理系陈庆教师、数学系陈嘉琼教师,他们的教学和科研工作,成绩都很好。

湖南大学是湖南省区现有高等学校四十七所之一,1934年我就来到了这里执教,至今已二十六年了,工作在这里,生活在这里。把解放前的十五年和解放后的十一年两相对比,真是两个世界。初来湖大,每个礼拜上了十几小时的课以外,什么也不管。为了生活,不得不在一个建筑公司兼任工程师。抗战八年,过着流浪生活。胜利后的几年里,生活更是苦闷。十五年的时间,哪里有心思去为教学,更谈不上什么学术成就和社会贡献了。新中国成立后,看到湖南大学的新生,我也变得年轻了,特别是党对我的信任,使我深深感觉到只有在新中国新的社会制度下,我们这些老教师老科学技术工作者才有发挥作用的机会。十一年来,国家一直给我以重任,在校内我担任一部分行政领导工作,一部分教学和科学研究工作,在校外我担任湖南省人民委员会委员、中国人民政治协商会议委员,我身心愉快,工作愈来愈有劲。我的家庭也是非常幸福的,七个儿女,六个已由国家培养从大学毕业出来,走上了各自的工作岗位,他们是国家的干部、医师、技师、研究员、教师,分布在北京、内蒙古、兰州、长沙、列宁格勒等地。家中只留着一个高中二年级的孩子和我的爱人,我们都生活得很舒适。我在工作之余,还常到全国各地去参观访问,看到祖国社会主义建设的辉煌成就,全国人民美好生活的景象,真使我兴奋!我虽年近古稀,而体力如昔,精神焕发,我要为26年朝于斯夕于斯的湖南大学而努力,我要把毕生精力投入到社会主义建设事业中去! (柳士英 写于1960年)

后记·致谢　Epilogue & Acknowledgement

　　湖南大学校园建筑的发展经历了岳麓书院时期、1920-1950年代开放理性的营造时期、改革开放后新地域校园的建构时期，每个时期校园规划都发生着相应的变化，整体空间随着湖南大学的发展是自然生长的，是有机扩张的，是呈现历时性同构的。在校园空间整体格局中，一代代建筑师精心设计建造的校园建筑无疑是重要的载体，而湖南大学校园也成为我国少有的具有十余栋国家重点保护文物建筑的高校。

　　The campus architecture of Hunan University has experienced three stages of transformation evolving from the archaic academy, to rationalist modern architecture during 1920-1950, and been gradually becoming a new regionalist campus since 1978. The changing pattern of the campus architecture over these years passing by explicates its nature of development: diachronic evolution within isomorphic context. The very buildings composing the campus had been meticulously designed by the influential architects of their time through generations and therefore become exceptional embodiments to indicate its growth and extension formation at the same time. It is well-known that this university campus is also the rare case having ten neoteric protect buildings of China.

　　对湖南大学校园建筑研究后发现，湖大校园的发展有三个特定的内核支撑，一是文脉传承。大学可以通过有形的校园建筑传承文脉，岳麓书院是原点，湖南大学的校园以其为心展开，形成层级结构，各个年代的建筑均以书院场所和人文精神塑造其精神内核，不论规模扩容、功能变异、风格变化，但这个内核始终是统一的。二是自然环境。湖大校园位于长沙山水洲城城市格局核心地带，岳麓山自古人杰地灵，成为国家5A级风景名胜区后，对校园发展中与自然环境的关系提出了更高的要求。校园近年来的发展逐步沿岳麓山开始东延至两山一湖区域（天马山、凤凰山和桃子湖），南延至后湖区域，校园与山体、湘江、湖泊、大树犬牙交错，浑然一体。因此在校园建筑的设计中也始终需要思考与这些自然要素的关系。三是城市型大学。湖南大学秉承现代大学开放、包容之办学精神，学科门类齐全，力争建设国家双一流综合性大学。其校园空间也始终是开放的（不设校门），与城市是融为一体的（景区、居民、公交线路融汇其中），是适应当代高水平大学办学需求的（岳麓山大学城的资源共享）。几十年来，湖南大学的校园建筑交织着文脉的、自然的、城市的这三个特征，从书院到天马校区的建筑均以此为基础不断延续完善，每一个历史时期的建筑都有它的特定性和它所在区域的合理性。

　　It is found that the underlying principles guiding the way of the campus redevelopment in the modern time are as follows: 1) Context awareness and culture inheritance. The legacy of Yuelu academy is the spirit of erudite and humanity passing down to new generations through the vicissitude of campus architecture, and it was although now still to rely on the tangible objects, for instance, campus buildings, to carry on these genes. Meanwhile, the spatial structure of the campus development reflects this identity by creating hierarchy pattern with respect to the past, placing the academy at the center place; 2) Environmental harmony. The location of the Hunan University campus gives it unique possibilities and challenges, since Yuelu Mountain at the west of the campus is listed as Five-star scenic resource of the Changsha city and the core urban environment element. At the mean time the Tianma valley at the south, Phoenix hill at the north side and Taozi Lake lying at the east of the campus all stretch north-southwards, which results in the same pattern of the campus extension ending at the Houhu lake at the south side. These integral nature settings provide prime condition for research discussion and study in the Hunan University, and hence have been considered as a whole over years of campus redevelopment;3) Urban university and openness. Although going through a few transformation of development, planning of the campus places emphasizes at all time on culture context, harmony with nature, openness to city following the university-running tenet of "open mind with inclusive manner". The goal of the university to be the world-class one makes it cooperating actively with other universities in the neighborhood of Yuelu district and striving to become the most open educational place and sharing space for the public of the city, therefore to accommodate the needs for the modernized urban university. To sum up, this urban campus generally features pluralistic culture and picturesque scenes and as results of redevelopment over years, it emerges 'Heterogeneous' forms in the 'Isomorphism' context in the sense of rationality.

2013年，我们曾搜集湖南大学重要建筑的原始图纸、各角度照片、相关背景资料，编著《异质同构——从岳麓书院到湖南大学》一书。适逢2019年，刘敦桢、柳士英先生创办湖南大学建筑学科90周年之际，湖南大学的校园建筑在近几年又增添了几栋新的重要建筑，我们再次思考了湖大校园建筑历史脉络的延续特征，重新搜集梳理原书资料，撰写学术论文，增加新建筑的介绍，更改原书开本，著得新书以纪念刘柳二位先生创办湖大建筑学科90周年。

After years of compiling the relevant materials and information, the Book "Heterogeneous Isomorphism" was published in 2013. Whereas it is the year of 2019 that the discipline of Architecture established in Hunan University is to celebrate its 90 anniversary, and newly built buildings emerge their significance for the campus development in the recent years. Those give rise to acts of re-writing and renewing the Book along with its overall format so as to commemorate such important moments both for the School of Architecture and Hunan University.

　　在书籍写作和编辑过程中，学院柳肃、袁朝晖、卢健松老师参与了多次关于写作内容与写作素材的会议，许昊皓、王蔚搜集了部分新增的图片和改绘，吴炳宇、许逸伦、于思璐、刘书惠、赵茂繁、林朝晖等同学绘制了校园老建筑线稿图，胡骉、姚力拍摄了部分新增照片，杨振航、廖靓雯提供了部分新建筑的资料，罗漾重新编排了全书版式，谢菲进行了英文翻译和校对工作。正是这些老师和同学们为新书提供了很好的建议，为本书的出版提供了很多珍贵的资料，在此一并致谢。

These aforementioned works accomplished should sincerely thank to the following inputs, including Mr. Liu Su, Mr. Yuan Zhaohui and Mr. Lu Jiansong's patiently drafting and re-drafting on the new contents, Mr. Xu Haohao and Mr. Wang Wei's supports for some re-drawings and additional photos, Wu Binyu, Xu Yilun, Yu Silu, Liu Shuhui, Zhao Maofan and Lin Zhaohui on drawing some of the historical architecture, Mr. Hu Biao and Mr. Yao Li's photographs contribution on a few more buildings, Mr. Yang Zhenhang and Ms. Liao Liangwen's partial provision for the new building information, Mr. Luo Yang's efforts on the format re-arrangement as well as Ms. Xie Fei's English translation and proofreading.

　　大学校园的历史感以及文化氛围，一如台阶上的青苔，必须一点点长出来，不可能一蹴而就。湖南大学不仅有着历经千年的岳麓书院，也有以柳士英先生为代表的前辈留下的许多经典建筑。沿着刘敦桢、柳士英先生"营造"之路，改革开放后的新地域建构时期仍然遵循着这种脉络关系，"承朱张之绪，取欧美之长"，继续"建构"校园空间和建筑。从营造到建构，异质而同构。

To create a university campus integrated with culture context, one should only count on efforts of a long period. Whether it is the Yuelu Academy established a thousand years ago, or architectural works done by those predecessors, i.e. Mr. Liu Dunzhen and Mr. Liu Shiying, or contemporary campus buildings built in the recent years, they all comply with the principle of "Harmony in diversity", or "Heterogeneous isomorphism" which is another term for it. Specifically speaking, creating a university campus of heterogeneous isomorphism is to fully understand the regional culture and environment upon which a new regional architecture could be built, to respond to the varying conditions by innovative approaches and eventually to attain "harmony in diversity" in a sustainable way.